软件自动化测试开发

邹辉 编著

电子工业出版社
Publishing House of Electronics Industry
北京·BEIJING

内 容 简 介

本书全面深入地介绍了软件自动化测试开发领域方方面面的相关知识，包括 App 功能自动化测试的方案、环境、代码运行及报告（基于 Appium 工具和 Java 语言编写），API 接口自动化测试的方案、环境、代码运行及报告（基于 Zentao 工具和 Python 脚本语言编写），Selenium 的 Web 自动化测试（基于 Selenium 工具和 Java 语言编写），JMeter 接口和性能测试，LoadRunner 性能测试以及 Jenkins 持续集成。

本书适用于想要了解、学习和使用当前流行的自动化测试开发技术的广大开发和测试从业人员，以及产品测试开发 leader 等。

未经许可，不得以任何方式复制或抄袭本书之部分或全部内容。
版权所有，侵权必究。

图书在版编目（CIP）数据

软件自动化测试开发 / 邹辉编著. —北京：电子工业出版社，2017.1
ISBN 978-7-121-30453-8

Ⅰ．①软… Ⅱ．①邹… Ⅲ．①软件－测试－自动化②软件开发 Ⅳ．①TP311.5

中国版本图书馆 CIP 数据核字(2016)第 283010 号

责任编辑：安　娜
印　　刷：三河市鑫金马印装有限公司
装　　订：三河市鑫金马印装有限公司
出版发行：电子工业出版社
　　　　　北京市海淀区万寿路 173 信箱　邮编 100036
开　　本：787×980　1/16　印张：17　字数：258 千字
版　　次：2017 年 1 月第 1 版
印　　次：2018 年 4 月第 5 次印刷
定　　价：59.00 元

凡所购买电子工业出版社图书有缺损问题，请向购买书店调换。若书店售缺，请与本社发行部联系，联系及邮购电话：(010) 88254888，88258888。
质量投诉请发邮件至 zlts@phei.com.cn，盗版侵权举报请发邮件至 dbqq@phei.com.cn。
本书咨询联系方式：010-51260888-819，faq@phei.com.cn。

前　言

关于本书

互联网软件技术发展速度非常快，稍不留神就"out"了，然而测试思路、开发语言却是有历史承传的。所以我们只有多实践打好基础，知其然并知其所以然，方可以不变应万变。我们上学也是从幼儿园、小学、中学……到博士导师一级一级上，学习技术也不例外，因此同时具备快速学习的能力也十分重要。人们对美好先进的东西永不会满足，追求永无止境，人生真的就是一个永远自强不息的成长过程。

本书用详细、完整的案例，完整的代码以及理论知识引领读者进入实际操作。当读者实践一遍之后，就能看到真实效果，并且印象深刻。就像拿到驾照后终身不用再考，而且能够驾驶准驾车型的各种车一样，二者的原理相同。这样我们就能花最小的代价学到最实实在在的自动化测试开发技术。

本书全面深入地讲解了自动化测试开发技术，包括接口自动化测试、App 自动化测试、Web 自动化测试和性能自动化测试，讲解了目前最新、最流行的自动化技术：移动 Appium 功能自动化、API 后台接口自动化、LoadRunner 性能测试、Selenium 的 Web 自动化以及 JMeter 自动化技术等。

本书基于 Java 语言和 Python 语言编写，结合各种主流开源工具框架而开发，能够真正地提升测试技术水平。学习测试开发掌握一门开发语言是极为必要的，读者可着重学习书中完整的 Java 和 Python 源代码和解析，提升薄弱环节。源代码对于实战来说十分重要，因此书中代码部分都加上了注释和解析，以供读者理解和学习。写代码无疑是测试同行较为薄弱的部分，但是当把看代码、写代码当成一种习惯或一种乐趣时，学习编码自然就能得心应手。

本书实践与理论相结合，包括方案、环境、代码和运行报告。书中的源码在保留版权的情况下供读者使用，也就是说，读者使用源码时需要注明出自本书。

本书需要读者有较好的基础和耐心，以及领悟力。对初学者而言，可以照葫芦画瓢，在实践中激发兴趣和信心，对有基础的读者而言，可以更加深入地了解自动化测试，并直接应用到工作当中，本书的内容将起到一个实用指南的作用。有的人擅于自己摸索着学习，有的人喜欢在书本的引导下学习，也有的人需要通过培训来学习，这都没关系，俗话说：不管黑猫白猫，抓到老鼠就是好猫。

关于软件自动化测试开发

除手工测试外，其他都是自动化测试。因此，软件自动化测试开发指的是测试技术和开发技术相结合，用手动测试之外的测试技术，模拟手动用户场景测试的实现过程，简单来说，就是用写的代码来测试程序。

在工作中，手动功能测试人员通常会感觉自己比其他如开发、产品等人员相对弱势，因为功能测试一般被认为技术含量不高，准入条件较低，甚至被认为工作就是单击页面，单击按钮。通常手动功能测试人员只需 3~6 个月左右的专业培训就足以胜任工作。

自动化测试开发工作相对手动功能测试而言，技术含量较高，相关人员的待遇也相对较高，他们与广大程序员的性质相同，是测试从业人员追求技术进步的表现。一般来说，自动化测试开发需要 1~3 年左右的专业培养方能胜任工作。

适合读者

绝大多数适用的读者

- 所有软件测试从业人员，包括测试团队 leader。
- 有一定开发语言基础的测试人员。
- 软件测试专业的学霸。

少部分适用的读者
- 软件测试专业的在校大学生。
- 软件技术、移动互联网相关人员，包括开发人员，研发团队 leader 等。
- 其他任何对软件自动化测试开发感兴趣的人。

作者简介

本书作者有丰富的软件测试从业经验，擅长开发和测试技术，担任测试主管，以电子商务、银行证券、移动互联网为主要业务领域，现在负责一上市企业 O2O 产品的软件自动化测试开发工作。

大纲内容

第 1 章到第 4 章：介绍 App 功能自动化测试方案、环境、代码、运行报告。

第 5 章到第 7 章：介绍 API 接口自动化测试方案、环境、代码、运行报告。

第 8 章：介绍 Selenium 的 Web 自动化测试。

第 9 章：介绍 JMeter 接口和性能测试。

第 10 章：介绍 LoadRunner 性能测试。

第 11 章：介绍 Jenkins 持续集成。

附录 A-D：小知识参考。

前言后记："鸡汤"。

致谢

感谢之前相对比较默契的同事，因为和大家共同的工作经历和成长才得来本书各个章节精彩内容的酝酿——分别有测试经理蔡祥；走秀网 PM 吴盛幸，开发劳水生，测试经理曾春刚；腾讯开发郑双明；华南城华盛电子商务 CTO 阳志军等。

最最感谢的是对本书感兴趣的你——本书的读者。

关于勘误

虽然书中的每个技术点都曾在实际项目中实践和应用过,但也会因为我们个人技术、所测项目和视野的局限,以及本人因时间仓促和能力水平等种种原因,书中难免会有一些错误和纰漏,如果大家在阅读过程中发现了什么问题,恳请反馈给我,读者朋友们可即时在线交流,联系方式如下,更多精彩内容,请关注微信公众号:

作者微信和 QQ 号:zouhui1003it,7980068

测试博客:http://www.cnblogs.com/finer

读者实战 QQ 互动群:377029807

微信公众号

更多精彩内容,请关注微信公众号:测试开发社区

微信扫一扫即可关注

目 录

第 1 章 App 自动化测试方案 ... 1
- 1.1 概述 ... 2
- 1.2 风险分析 ... 2
- 1.3 软硬件需求 ... 3
- 1.4 测试计划 ... 3
- 1.5 Appium 移动自动化框架 ... 4
- 1.6 测试框架 ... 10
- 1.7 自动编译部署工具 ... 11

第 2 章 Android 自动化环境搭建 ... 14
- 2.1 Android 搭建的简要步骤 ... 15
- 2.2 在 Windows 上搭建 Android 自动化环境 ... 16
- 2.3 在 Mac 机器上搭建 Android 自动化环境 ... 28
- 2.4 Android 自动化测试运行 ... 29

第 3 章 iOS 自动化环境搭建 ... 30
- 3.1 iOS 环境搭建的简要步骤 ... 31
- 3.2 iOS 自动化环境搭建的详细步骤 ... 31
- 3.3 iOS 自动化测试运行 ... 38
- 3.4 iOS 的 App 自动化测试 demo 演示视频 ... 38

第 4 章 App 自动化测试源代码 ... 39
- 4.1 基于 Java 的 App 自动化源代码解析 ... 40
- 4.2 源代码结合 Ant 持续集成到 Jenkins ... 71
- 4.3 Android 和 iOS 自动化测试结果展示 ... 73

第 5 章 API 接口自动化测试方案 ... 75
- 5.1 概述 ... 76
- 5.2 所用技术点 ... 78
- 5.3 主要功能 ... 78
- 5.4 测试计划 ... 79

第 6 章 API 接口自动化环境搭建 ... 80
- 6.1 Python 环境准备 ... 81
- 6.2 Zentao（禅道）项目管理工具 ... 83
- 6.3 MySQL 数据库 ... 84
- 6.4 Fiddler 接口抓包工具 ... 86
- 6.5 Postman 接口测试工具 ... 93

第 7 章 API 接口自动化源代码 ... 96
- 7.1 基于 Python 的接口自动化脚本解析 ... 97
- 7.2 Python 接口测试数据展示 ... 147
- 7.3 脚本持续集成到 Jenkins ... 151
- 7.4 接口自动化测试报告 ... 151

第 8 章 Selenium 的 Web 自动化测试 ... 154
- 8.1 Selenium 自动化测试准备 ... 155
- 8.2 Selenium 自动化源码解析 ... 156
- 8.3 持续集成到 Jenkins ... 173
- 8.4 Web 自动化测试结果展示 ... 175

第 9 章 JMeter 接口测试和性能测试 ... 177
- 9.1 安装和介绍 ... 178
- 9.2 Jmeter 接口测试示例 ... 190

9.3　结合 Ant 持续集成到 Jenkins 196
9.4　接口测试结果 199
9.5　JMeter 性能测试示例 200

第 10 章　LoadRunner 性能测试 204
10.1　小概念 205
10.2　安装 214
10.3　脚本调试 217
10.4　运行场景 221
10.5　性能监控 223
10.6　问题分析和调优 224
10.7　性能压力测试报告样例 226

第 11 章　Jenkins 持续集成 232
11.1　介绍 233
11.2　系统配置 233
11.3　项目配置 236
11.4　多机器节点配置 240
11.5　结果展示视图 243

附录 A　自动化管理平台和产品自动化系统 245

附录 B　Java 和 Python 开发语言学习历程 247

附录 C　常见错误和问题解答 252

附录 D　常用软件安装包链接 255

后记 259

第 1 章

App 自动化测试方案

1.1 概述

什么是 App 自动化？为什么要做 App 自动化？

App 自动化是指给 Android 或 iOS 上的软件应用程序做的自动化测试。

手工测试和自动化测试的对比如下。

手工测试优势：不可替代、能发现更多 bug、包含了人的想象力与理解力。

自动化测试优势：可重复、效率高，能增加对软件质量的信任度。

> 注意，不是所有功能都需要自动化，只需把重复执行的以及主要的交给自动化。

App 自动化测试的特点如下：

- 执行自动化测试只能发现一小部分 bug。
- 执行自动化冒烟测试或回归测试是用来验证系统状态，而不是找出更多 bug。
- 执行自动化测试可以让测试同事有更多的精力来关注复杂场景，做更多更深层次的测试。
- 编写自动化测试过程中会发现一部分 bug，发现后要及时记录。

1.2 风险分析

自动化测试的主要风险分析如下：

（1）测试用例覆盖率（覆盖率决定了测试效率，因此要选择合适的用例，应约占功能用例集的 20%~50%）。

（2）测试结果准确度（准确度决定了测试有效性，因此应尽可能减少误报）。

（3）自动化代码维护（维护影响成本，本书写的是关键字驱动自动化框架，自动化框架代码应尽可能优化，所测的功能改动而代码不需要改动时才是强大的框架，维护成本才足够低）。

（4）版本开发和测试时间进度（当项目需求和功能较为稳定时，建议用自动化测试）。

（5）开发对控件元素增修改的程度（需开发人员尽可能地用 name 元素，并且和 UI 设计一致。当修改变动量较小时，测试人员可根据提供的元素提前介入，开发自动化脚本）。

App 源码权限控制，在 iOS 上测试时需要用到源码，我们测试人员可能只需要 SVN 下载权限，不需要上传权限，因此应尽可能地避免改动 SVN 上开发人员的源码。

1.3 软硬件需求

自动化测试的软硬件需求如下。

硬件：

- Mac 电脑、iPhone 手机。
- Windows 电脑、Android 手机。

软件：

- Appium 测试框架：运行 App 驱动的自动化平台，通过识别的控件元素，模拟用户的手工操作，支持 iOS 和 Android 系统。
- Jenkins：持续集成自动构建和执行任务。
- TestNG、Ant、SVN：测试插件初始化、测试、断言、清理。
- JDK、Eclipse、Java 语言开发编写环境。
- AdbWireless：安卓手机和电脑间的无线连接。

1.4 测试计划

时间计划

对于有良好代码基础的熟手，可用一周时间做出演示 demo。如果是从零开始的小白，则可用 3 到 6 个月的时间做出演示 demo。

对于有良好代码基础的熟手，可用一个月时间试运行冒烟测试用例。如果是从零开始的小白，则可用半年到一年的时间试运行测试冒烟测试用例。

目前 App 自动化框架计划

采用自动化关键字数据驱动模式设计，即表格驱动测试或者基于动作的测试。关键字驱动框架的基本工作是将测试用例分成四个部分：一是测试步骤，二是测试步骤中的对象，三是测试对象执行的动作（Action），四是测试对象需要的数据（Test Data）。

后期 App 自动化框架计划

把测试用例、控件元素等放入数据库，前端页面进行展示和操作。做到写自动化测试用例完全不用增修改代码，自由管理大量用例和测试数据，最终做成自动化平台。关于自动化测试平台的开发实现，可阅读我的另一本书《自动化平台测试开发》—是基于 Python 语言编写，算是本书的高阶版，主要功能可参见本书附录 A。

1.5　Appium 移动自动化框架

1. 需要要掌握的技能（第 1 章到第 4 章都会用到）

（1）Appium API、WebDriver 基础知识和环境搭建（见第 1.5 节）。

（2）TestNG 等测试框架（见第 1.6 节）。

（3）Android/iOS 开发测试基础以及环境搭建（见第 2 章和第 3 章）。

（4）开发移动自动化项目的 Java 语言或 Python 语言等（见第 4 章）。

需要说明的是，如果想用 Python 语言编写 App 自动化测试框架，那么读者可以参考阅读我的另一本书《自动化平台测试开发》—是基于 Python 语言编写。Appium 自动化测试框架的功能概括如下：

2. Appium 框架的功能

（1）支持 iOS、Android，可在多台机器上并行 App 自动化，测试机型适配。

（2）代码实现关键字驱动：

- 测试集：关联 Excel 测试用例和脚本配置。
- 测试数据：Excel 存储输入数据、控件元素、测试结果。
- 测试脚本：由 Java 和 TestNG 编写，分层结构有 case、log、config、report 以及 data 等。

（3）自动测试用例执行：

- 从功能测试用例中抽取需重复执行的、主要的功能进行用例覆盖。
- 支持用例 failed（失败）时自动截屏。
- failed（失败）用例自动重复执行数遍。

（4）持续集成环境 Jenkins，定时自动构建和执行测试任务。

- 测试结果报告展示，自动邮件展示。

Appium 自动化测试一个 App 的基本过程如下：

3．测试 App 的基本过程

基于 Appium 自动化测试框架，我们要进行的是连接电脑、连接手机、解锁、安装 App、卸载 App、启动 App、元素定位、元素的操作、屏幕的操作、页面等待、异常处理截图、数据校验、日志、报告等一系列自动化测试执行的详细过程。

Appium 自动化框架元素控件的捕获，根据捕获到的元素控件进行相应的操作。

Appium 元素控件有多种定位方法，最常用的是元素的 ID（即 By.id）和元素的值（即 By.name）。还可以通过元素类型 TagName、元素的位置 XPath、手机设备的坐标等进行定位操作。安卓的元素控件可以通过 SDK 中的 uiautomatorviewer.bat 文件进行录制和捕获定位，如图 1.1 至图 1.3 所示。

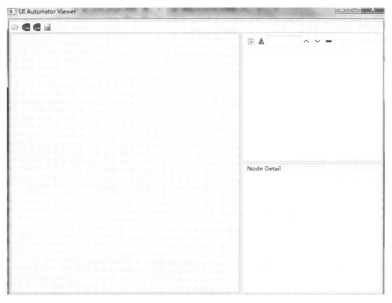

▲图 1.1

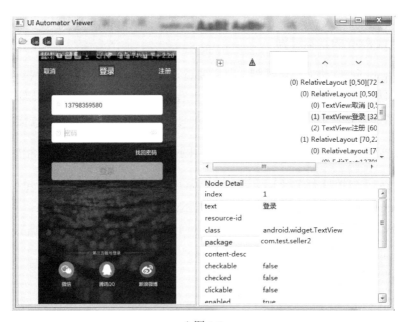

▲图 1.2

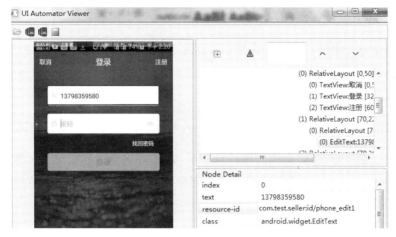

▲图 1.3

如图 1.3 所示，Node Detail 下面的 resource-id com.test.seller:id/phone_edit1 对应 Excel 和代码中的定位方法 By.id，控件元素数据 text 13798359580 对应 Excel 和代码中的操作方法 sendkeys()，控件元素赋值数据为 13798359580。

可以这样理解：首先找到这个文本框，接着给这个文本框输入数据。即通过 ID 属性值 com.test.seller:id/phone_edit1，找到此用户名文本框的控件元素，然后通过 sendkeys()方法输入用户名数据 13798359580 到此用户名文本。其他自动化测试步骤的定位方法、控件元素以及操作方法也都与此类似。实际上，自动化测试就是通过程序代码来实现模拟手动测试去操作一遍的过程。

上面介绍了用户名文本框输入用 sendkeys()方法，那么其他元素的操作方法有哪些呢？元素操作方法大致有单击（click）、输入（sendkeys）、元素滑动、页面滑动、长按、下拉、弹出、屏幕放大缩小等，最常用的就是单击和输入。代码解析详见第 4 章中的源码。

数据校验。其实元素本身就是数据校验，当程序找不到元素时，用例就会失败。另外，测试用例时可以加入一个或多个断言进行验证数据，还可设置步骤等待延迟时间，详细内容见第 4 章讲解中的源码。

测试结果。测试用例中记录了运行后的测试结果，如 pass、failed 或是 skip，详细内容见第 4 章关联的 Excel 测试用例。

4．Appium 介绍（参考 Appium 官方资料）

Appium 是一个移动端自动化测试开源工具，支持 iOS 和 Android 平台，支持 Python、Java 等语言，即同一套 Java 或 Python 脚本可以同时运行在 iOS 和 Android 平台。

Appium 是跨平台的，即可以针对不同的平台用一套 API 来编写测试用例。

Appium 是一个 C/S 架构，核心是一个 Web 服务器，它提供了一套 REST 的接口。当收到客户端的连接后，就会监听到命令，然后在移动设备上执行这些命令，最后将执行结果放在 HTTP 响应中返还给客户端。

5．Session

自动化始终围绕一个 Session（会话）进行。客户端初始化一个 Session 来与服务端交互，不同的语言有不同的实现方式，但是它们最终都是发送一个 POST 请求给服务端，请求中包含一个 JSON 对象，其被称作"Desired Capabilities"。此时，服务端就会开启一个自动化的 Session，然后返回一个 Session ID，Session ID 将会被用户发送后续的命令。

6．Desired Capabilities

Desired Capabilities 是一些键值对的集合（比如一个 map 或者 hash）。客户端将这些键值对发送给服务端，告诉服务端我们想要怎样测试。比如，我们可以把 platformName capability 设置为 iOS，告诉 Appium 服务端，我们想要一个 iOS 的 session，而不是一个 Android 的 session。

7．Appium Server 服务端

Appium Server 是用 Node.js 写的，我们既可以用源码编译，也可以从 NPM 直接安装。

Appium 服务端有很多语言库，如 Java、Ruby、Python、PHP、JavaScript 以及 C#等，这些库都实现了 Appium 对 WebDriver 协议的扩展。当使用 Appium 的时候，你只需使用这些库代替常规的 WebDriver 库就可以了。

8. Appium Clients 客户端

此客户端的概念不是我们传统意义上的客户端，更好的理解方式是一个扩展的 WebDriver 协议库，当你用自己喜欢的语言写 case 时，会将该语言扩展的 WebDrvier 库添加到自己的环境中，这时你可以把它理解为这就是个客户端。

Appium Clients 客户端的安装包如下。

Mac 机器上直接运行 Appium.dmg；Windows 机器上运行 Appium.exe。

9. Appium Android/iOS 工作原理

API 接口调用 Selenium 的接口，Appium Server 接收 WebDriver 标准请求，解析请求内容，调用对应的框架响应操作。代码将 DesiredCapability 中的键值对组合成一个 JSON，然后通过 HTTP 协议发送到 Appium 服务器创建一个 session。代码与 Appium 的所有交互都是围绕着这个 session 进行的。session 创建成功后，Appium 再通过 USB 接口与手机之间创建 TCP 连接，先安装一些服务端 App，比如 Android API 4.2+是 uiautomator，Android 2.3+是 Instrumentation；如果是 iOS，则是 UiAutomation。手机的操作都是由 Appium 发送指令到 uiautomator，然后再由 uiautomator 进行控制的。

Appium 原理图如图 1.4 所示。

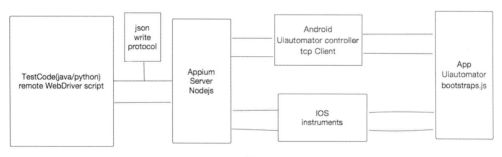

▲图 1.4

Appium 的核心是一个遵守 REST 设计风格的 Web 服务器，它接收客户端的连接和命令，在手机设备上执行命令，然后通过 HTTP 的响应收集命令执行的结果。这种架构给我们提供了很好的开放特性：只要某种语言有 HTTP 客户端的 API，我们就可以通过这个语言写自己的测试代码。

1.6 测试框架

TestNG 简介（参考 TestNG 官方资料）

TestNG 是一款基于 Java 的测试框架，被设计用于解决大部分的测试需求，涵盖单元测试（测试一个单独的类）和集成测试（测试由几个类、几个包甚至有几个框架组成的系统）两种测试方式。

一般情况下，一个测试通常需要以下三步：

（1）写出你需要测试的业务逻辑，并在你的代码中加上 TestNG 注解。

（2）在 TestNG.xml 或 build.xml 中加上测试信息（例如，你需要运行的类名、组名等）。

（3）运行测试。

一个 suite 使用一个 XML 文件来定义。该 suite 可以包含一个或多个 test，且该 suite 使用<suite>标签来定义。

一个 test 使用<test>标签来定义，该 test 可以包含一个或多个 TestNG 类。

一个 TestNG 类就是一个 Java 类，它至少包含一个 TestNG 注解。一个 TestNG 类使用<class>标签来定义，可以包含一个或多个方法。

一个测试方法就是一个在你的代码中使用@test 注解标注的 Java 方法。

一个 TestNG 测试可以使用@BeforeXXX 或@AfterXXX 注解，并被配置为用于在某一切入点之前或之后执行一些代码逻辑，这些切入点可以是上述所列项中的任意一项。

下面简要地介绍一下 TestNG 中的注解进行。

- @BeforSuite： 被标注的方法将在本 Suite 中的所有测试运行之前运行。
- @AfterSuite： 被标注的方法将在本 Suite 中的所有测试运行之后运行。
- @BeforeTest： 被标注的方法将在本测试运行之前运行。
- @AfterTest： 被标注的方法将在本测试运行之后运行。

- @BeforeGroups： 被标注的方法将在本 Groups 中的所有测试运行之前运行。
- @AfterGroups： 被标注的方法将在本 Groups 中的所有测试运行之后运行。
- @BeforeClass： 被标注的方法将在本 Class 中的所有方法执行之前运行。
- @AfterClass： 被标注的方法将在本 Class 中的所有方法执行之后运行。
- @BeforeMethod： 被标注的方法将在每一个测试方法前执行。
- @AfterMethod： 被标注的方法将在每一个测试方法后执行。

你可以使用多种不同的方式运行 TestNG，例如，可以使用 testing.xml 文件、Ant 或命令行。

你可以在 testng.xml 内部定义新的组，并且可以在属性中增加其他信息例如，是否平行运行测试，使用了多少线程，是否运行 JUnit 测试等。

另外，TestNG 的详细介绍还包括执行测试、嵌套测试、忽略测试、组测试、异常测试、依赖测试、参数化测试，以及测试结果报告等，读者可查找并参考 TestNG 官方的相关文档资料。

1.7 自动编译部署工具

本节内容本应在第 11 章结合 Jenkins 学习，但由于第 3 章代码中涉及 Ant 的内容，所以提前介绍一下，读者也可以跳过本节，等环境搭建好，demo 运行起来后再来学习这部分内容。

1．Ant 简介（参考 Ant 官方资料）：

一：Ant 是一个将软件编译、测试、部署等步骤联系在一起加以自动化的一个工具，一般集成到 Jenkins 中，多用于 Java 环境中的软件开发。在实际软件开发中，有很多地方都可以用到 Ant。Ant 的升级版是 Maven，大家也可以使用 Maven 来代替 Ant。

Ant 是 Apache 软件基金会 JAKARTA 目录中的一个子项目，它的优点如下。

（1）跨平台性：Ant 是由纯 Java 语言编写的，所以具有很好的跨平台性。

（2）操作简单：Ant 由一个内置任务和可选任务组成，用 Ant 任务就像在 DOS 中写命令行一样。Ant 通过调用 target 树，可以执行各种 task，每个 task 实现了特定接口对象。Ant 运行时需要一个 XML 文件（构建文件）。

（3）维护简单、可读性好、集成简单：由于 Ant 构建文件时用的是 XML 格式的文件，所以不仅易于维护和书写，而且结构十分清晰。由于 Ant 有跨平台性和操作简单的特点，因此它很容易集成到一些开发环境中去。

二：Ant 的构建文件，当开始一个新的项目时，首先应编写 Ant 构建文件。构建文件定义了构建过程，并被团队开发中的每个成员使用。Ant 构建文件默认名为 build.xml，也可以取其他的名字，只不过在运行的时候，需要把这个名字当作参数传给 Ant。构建文件可以放在任何位置，一般做法是放在项目的顶层目录中，这样可以保持项目的简洁和清晰。一个典型的项目层次结构如下所示：

（1）src 存放文件。

（2）class 存放编译后的文件。

（3）lib 存放第三方 JAR 包。

（4）dist 存放打包及发布后的代码。

Ant 构建文件是 XML 文件。每个构建文件定义一个唯一的项目（Project 元素），每个项目下可以定义很多目标（target 元素），这些目标之间可以有依赖关系。当执行这类目标时，需要执行它们所依赖的目标。每个目标中可以定义多个任务，目标中还定义了所要执行的任务序列。Ant 在构建目标时必须调用所定义的任务。任务定义了 Ant 实际执行的命令。Ant 中的任务可以分为三类：

（1）核心任务。核心任务是 Ant 自带的任务。

（2）可选任务。可选任务是来自第三方的任务，因此需要一个附加的 JAR 文件。

（3）用户自定义任务。用户自定义任务是用户自己开发的任务。

Ant 的配置文件为 build.xml，读者可结合本书第 11 章 Jenkins 持续集成时使用。

本书对 Ant 命令、Ant 脚本、标签、使用节点、元素和属性、命令指令、Ant 编译打包、运行工程等不做详细讲解，读者可自行查找官方资料。

2．Svn 简介（参考 Svn 官方资料）

SVN 是一个自由、开源的版本控制系统，是项目组人员存放在代码等文件的中心版本库，记录每一次文件和目录的修改。Subversion 允许把数据恢复到早期版本，或是检查数据修改的历史。Subversion 可以通过网络访问它的版本库，从而使用户可以在不同的电脑上进行操作。

我们把测试开发的源代码存放在 SVN 上，然后结合 Jenkins 进行持续集成构建时使用。

安装文件路径：http://subversion.apache.org/packages.html。

第 2 章

Android 自动化环境搭建

2.1 Android 搭建的简要步骤

Android 搭建的简要步骤如下：

1．安装 JDK，配置 JDK 环境变量。注意一下电脑是 32 位还是 64 位，需要与之对应。

2．安装 Eclipes 和火狐 Selenium IDE，注意一下电脑是 32 位还是 64 位，需要与之对应。

3．加入 jxl、log4j、Selenium、Gson 和 Java-Client 包。

4．加入 JUnit 包，创建 JUnit 测试类。

5．加入 TestNG 包，创建 TestNG 测试类 TestNG.xml。

6．安装 Ant，配置 Ant 环境变量 build.xml。

7．安装 Jenkins、Tomcat，配置 Tomcat 环境变量。

8．安装 Android-SDK-Windows。

9．安装 Appium。

10．配置 Android 环境变量。

Android 自动化几个关键点如下：

（1）抓取页面控件元素，使用工具 android-sdk-windows\tools>uiautomatorviewer.bat。

（2）安装 App 软件，使用工具 android-sdk-windows\platform-tools>adbinstall Aoaio.apk。

（3）获取 App 入口，使用方法 android-sdkwindows\tools aaptdump badging E:\apk\es3.apk，获取入口后写入代码配置文件中内容类似如下：

capabilities.setCapability("appPackage", "com.cs.aola");

capabilities.setCapability("appActivity", ".ui.StartActivity")

（4）安装 Android，使用工具 eclipseMacketplace。

（5）搭建自动化测试框架 Appium。

（6）在 Android 真机上测试时，由于模拟器适配，测试意义不大，因此不再赘述，感兴趣的读者可以研究一下。对于使用 android 对应的版本以及 android 的 API，Android 4.2 不支持 resources id，所以需要 Android 4.4 以上版本。

（7）root 安卓手机，在 cmd 中输入 adb devices，看看设备是否连接成功。

（8）无线连接运行 adbWireless;adb connect 192.168.XX.XX;，即可安装无线运行，由于手机必须 root，因此平时一般可以直接用数据线连，不用无线连接运行。

2.2 在 Windows 上搭建 Android 自动化环境

Appium 测试环境的搭建相对比较烦琐，相信不少初学者都花了很多时间在环境搭建上，所以本书对这部分内容单独进行讲解。Appium 服务端环境

1. Appium 服务端环境

（1）安装 Node.js

下载 Node.js 安装包（http://nodejs.org/download/），选择最新版本安装，如图 2.1 所示。

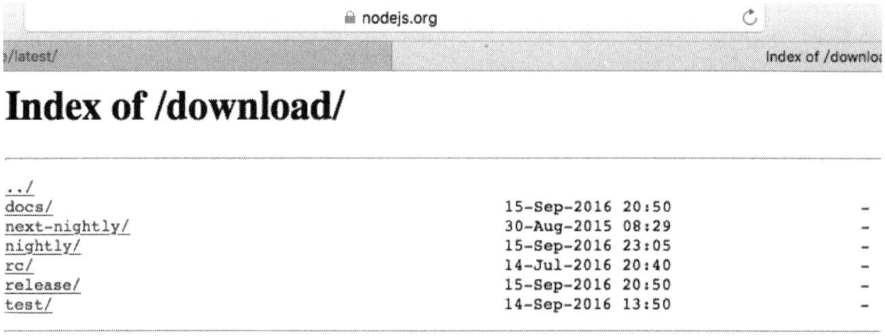

▲图 2.1

安装后，测试安装是否成功。运行 cmd，输入 node -v ，如图 2.2 所示。

▲图 2.2

安装文件路径：https://pan.baidu.com/s/1pKLwEFp 。

（2）安装 Android 的 SDK

安装 Android 的 SDK 包（http://developer.Android.com/SDK/index.html），运行依赖 SDK 中的"Android"工具并确保安装了 Level 17 或以上版本的 API。设置 ANDROID_HOME 系统变量为 Android SDK 路径，并把 tools 和 platform-tools 两个目录加到系统的 Path 路径中。

变量名：ANDROID_HOME

变量值：D:\android-sdk

变量名：Path

变量值：%ANDROID_HOME%\platform-tools;%ANDROID_HOME%\ tools;

安装文件路径：https://pan.baidu.com/s/1mi6PT9m

（3）安装 JDK

执行"下载→解压文件夹→安装操作"，并设置 Java 环境变量，执行"我的电脑→属性→高级→环境变量"命令，新建系统变量 JAVA_HOME 和 CLASSPATH。

变量名：JAVA_HOME

变量值：C:\Program Files (x86)\Java\jdk 1.7.0_01

变量名：Path

变量值：%JAVA_HOME%\bin;%JAVA_HOME%\jre\bin;

变量名：CLASSPATH

变量值：;%JAVA_HOME%\lib\dt.jar;%JAVA_HOME%\lib\tools.jar;

测试环境安装成功：运行 cmd，输入 java –version，如果成功则出现 Java 信息，如图 2.3 所示。

▲图 2.3

安装文件路径：https://pan.baidu.com/s/1gf4Ym3L。

（4）安装 Apache Ant

安装 Apache Ant（http://ant.apache.org/bindownload.cgi）。解压缩文件夹，并把路径加入环境变量。执行"我的电脑→属性→高级→环境变量"命令，新建系统变量 ANT_HOME。

变量名：ANT_HOME

变量值：D:\apache-ant-1.8.2

变量名：Path

变量值：%ANT_HOME%\bin

测试 Ant 环境是否安装成功。运行 cmd，输入 ant，如果没有指定 build.xml，则输出如图 2.4 所示。

▲图 2.4

运行 cmd，输入 ant –version，如果正常显示 Ant 的版本号，则说明 Ant 环境已搭建成功。

安装文件路径：https://pan.baidu.com/s/1c1IvthY。

（5）安装.Net 4.5

安装文件路径：https://pan.baidu.com/s/1sl1qdgL。

（6）配置并安装 Android 版的 Appium 客户端（http://appium.io/），并配置手机信息。如果是真机，则 Capabilities 部分的设置要填写准确。设置对应的 PlatformVersion 和 Device Name。

> 注意，Device Name 一定要与真机一致，既可以通过 adb devices 命令查找到后填写进去，也可以通过手机助手查看设备名称，然后填写进去。

运行代码前，单击图 2.5 中右上角的三角形▶图标，启动 AppiumPlatformVersion。比如选 4.3 时，SDK 一定要安装 Android 4.3.1（API 18）。

安装文件路径：https://pan.baidu.com/s/1jHGhnxG 。

▲图 2.5

（7）安装 Eclipse

设置 IDE 集成开发环境，注意 Windows 机器是 32 位还是 64 位，Eclipse 版本需与之对应。

安装文件路径：https://pan.baidu.com/s/1dF0sBcP。

（8）在 Eclipse 中安装 TestNG

执行"help→Install New SoftWare"操作，在 Eclipse 中 安装 TestNG 组件，路径为 http://beust.com/eclipse。

在安装 TestNG 组件时，因为有时访问 Google 需要"翻墙"，因此可以采用离线安装包进行安装。

安装方法：解压缩后覆盖 Eclipse 目录下相对应的文件或文件夹，注意路径一定要正确。

安装文件路径：https://pan.baidu.com/s/1bLhluA。

(9) 在 Eclipse 中安装 SVN 客户端

执行 "help→Install New SoftWare" 操作，在 Eclipse 中安装 sub 插件，路径为 http://subclipse.tigris.org/update_ 1.6.x。

安装后连接到服务端，比如 https://192.168.60.101/svn/。

SVN 简介：SVN 是代码版本控制管理工具，我们的源代码从 Eclipse 连接 SVN，最终存储到 SVN 服务端，然后结合 Jenkins 配置集成，自动读取并编译源代码。

(10) 在 Eclipse 中安装 Android 插件 ADT

执行 "help→Install New SoftWare" 操作，接着执行 "Eclipse 菜单中的 Help→install new software→add Archive; ADT-23.0.7.zip" 操作。

安装文件路径：https://pan.baidu.com/s/1sl2BZit。

(11).在 Eclipse 中配置 Android 的 SDK

执行 "Eclipse 菜单中的 Window→Preferences" 操作，进入如图 2.6 所示界面。

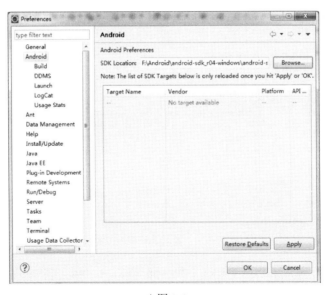

▲图 2.6

选择 Android SDK 解压后的目录，单击 Apply 按钮，单击"OK"按钮，安装文件路径：https://pan.baidu.com/s/1mi6PT9m。

（12）安装安卓手机相应版本的 API，单击图 2.7 中的 SDK 标志，弹出如图 2.8 所示界面，选中 Android5.1.1 和 5.0.1API 并安装它。

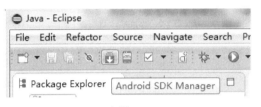

▲图 2.7

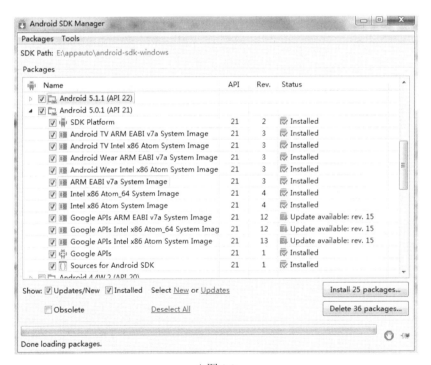

▲图 2.8

（13）查找所测 App 包名 Package 和 App 入口 Activity，可以通过以下方式：

- 请 Android 开发负责人提供。

- 到 sdk→tools 目录下的 dmms 查找日志中找。
- 运行 cmd，然后输入 adb shell，再输入 logcat *:S ActivityManager:V。

（14）查找安卓手机名 adb devices，或者用手机助手，可以看到手机已连接，如图 2.9 所示。

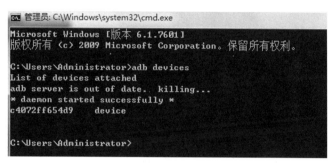

▲图 2.9

（15）新建名称，如 myproject 的工程项目，如图 2.10 所示。

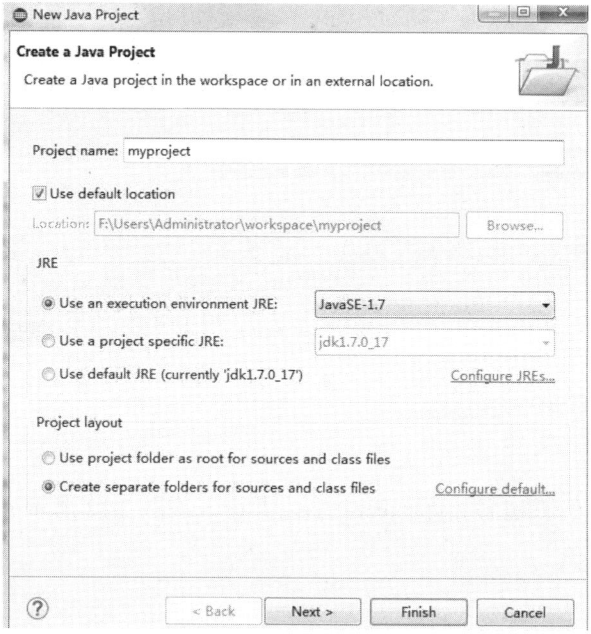

▲图 2.10

单击 Finish 按钮，如图 2.11 所示。

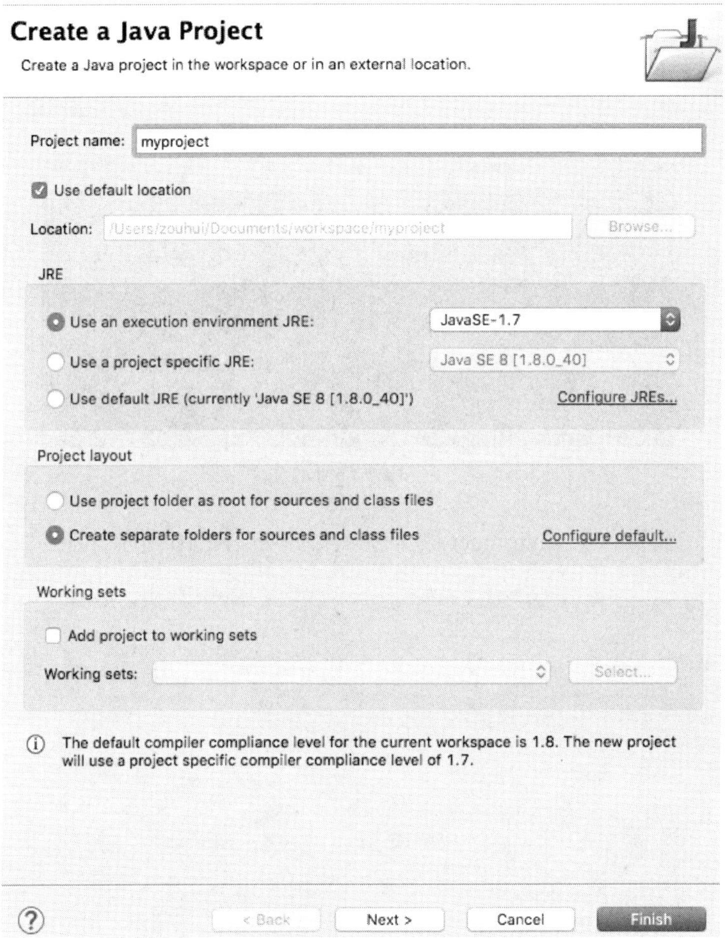

▲图 2.11

（16）导入项目文件

1）选择新建的工程名 myproject，右击并进入如图 2.12 所示界面。

第 2 章　Android 自动化环境搭建 | 25

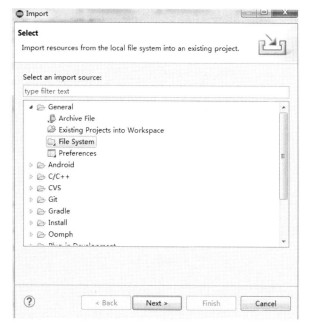

▲图 2.12

选择 File System，单击 Next 按钮，弹出如图 2.13 所示界面。

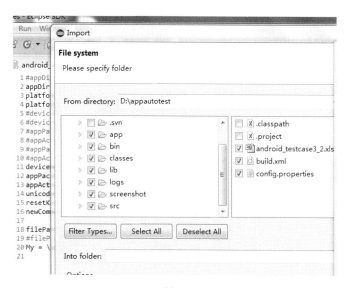

▲图 2.13

然后勾选所有文件集，但不勾选.svn、.classpath、.project，单击 Finish 按钮。

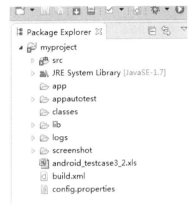

▲图 2.14

2）重新导入 lib 下的 jar 包。

导入到已建好的工程，并右击工程名，执行"Properti→Libraries→Add External JARs"操作，弹出 appautotestdemo 目录 lib 子目录下所有的 jar 包，如图 2.15 所示。

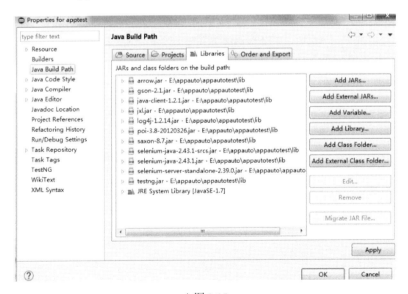

▲图 2.15

单击"OK"按钮，此时可以看到 myproject 工程文件正常，没有显示红叉。接下来改配置文件，Appium 手机信息与 App 信息一致就可以了。

当文件路径和原工程设置一致时，也可以直接导入工程文件，如图 2.16 所示，选第二项。

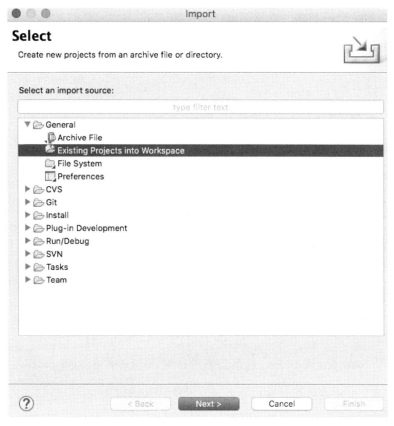

▲图 2.16

还可以按照第 4 章源代码，逐个新建文件并编写代码到 myproject 工程中。

接下来改配置文件，Appium 配置、手机信息与 App 信息一致且正确就可以了。运行时首先保证能启动所要测试的 App，然后再写登录 demo 用例，环境的搭建就完成了。

具体请阅读第 4 章源代码解析，主要有 android_config.properties、testng.xml、AutoLoginTest.java 文件以及 Appium 界面，还有 SDK 中的 API 等。

2.3 在 Mac 机器上搭建 Android 自动化环境

首先准备好 Mac 机器、安卓手机以及待测 App。

简要步骤参考如下，由于使用不常见，详细的本书就不介绍了：

（1）安装 JDK。

（2）安装 Eclipse。

（3）安装 SDK。

（4）安装 ADT。

（5）安装 Appium 客户端 appium.dmg。

（6）在 Mac 命令窗，安装 Appium 服务端，设置 Node.js 环境变量：Hello node 127.0.0.1：8000。

（7）在 Mac 命令窗，设置 SDK 环境变量步骤如下。

1）进入当前用户的 home 目录，创建文件：touch .bash_profile。

2）打开 .bash_profile 文件，对其内容进行编辑：open -e .bash_profile。

3）此时文本编辑器会打开一个文本，编辑加入以下内容：

    export   PATH=${PATH}:/Users/apple/Library/Android/sdk/platform-tools
    export   PATH=${PATH}:/Users/apple/Library/Android/sdk/tools

路径可根据 SDK 的安装目录进行修改（vi），保存文件（wq），关闭 .bash_profile。

4）执行生效，更新刚配置的环境变量，赋予 sudo 权限执行，命令如下：

    sudo source .bash_profile

5）验证：输入 adb 并回车。如果未显示 command not found，则环境变量设置生效。

（8）验证配置是否能成功连接 Android 手机，命令如下：adb devices。

（9）打开 Eclipse 新建项目，导入 App 自动化测试框架的源代码，进行验证。或者按照第 4 章源代码，逐个新建文件，并编写代码到 myproject 工程中。

2.4　Android 自动化测试运行

（1）连接手机，开启手机调试模式，查看与电脑连接是否正常。

（2）启动 Appium，无错误日志。

（3）在 Eclipse 中运行 TestNG 的代码脚本，可以看到在 Appium 客户端有值输出，在手机上也能看到自动化运行测试用例已经开始了。

（4）查看测试结果。

第 3 章

iOS 自动化环境搭建

3.1 iOS 环境搭建的简要步骤

1. 设置 JDK Java 环境变量。

2. 设置 Eclipse IDE 集成开发环境，把源代码导入到工程文件，参见第 6 章。

3. 安装 Xcode。

4. 安装 Node.js。

5. 搭建 Appium 自动化测试框架。

6. 设置 brew mac 环境。

7. 安装 wd。

8. sudo brew install --HEAD ideviceinstaller，手机版本设备安装。

9. iPhone 手机设置→开发者→UI Automation 开启。

10. 获取 SouceTree 版本，单击 3.1 Release 版本管理客户端。

11. 安装 Git。

12. 安装 Xcode command line tools。

13. 设计 Ant 编译、测试、部署自动化工具（基于 Java 使用）。

14. 设置 Jenkins（参见第 9 章，192.168.60.200:8080/Jenkins）。

15. 申请开发者调试证书。

3.2 iOS 自动化环境搭建的详细步骤

Appium 测试环境的搭建相对比较烦琐，不少初学者在此走过不少弯路，所以本书单独把它作为一章进行讲解。

首先是熟悉 Mac 电脑的使用习惯：命令行是 Linux 风格，界面则类似于 Windows

风格。

环境搭建步骤如下。

(1) 安装 JDK,并配置环境变量,方法如下。

sudo vi /etc/profile 在最后行加入如下内容:

```
JAVA_HOME="/Library/Java/JavaVirtualMachines/jdk1.8.0_40.jdk/Contents/Home/"
CLASS_PATH="$JAVA_HOME/lib"
PATH=".:$PATH:$JAVA_HOME/bin"
```

保存退出后,执行生效,输入命令 Source ~/.bash_profile。

输入命令 java –version,环境安装成功后如下所示:

```
zouhuideMacBook-Air:~ zouhui$ java -version
java version "1.8.0_40"
Java(TM) SE Runtime Environment (build 1.8.0_40-b27)
Java HotSpot(TM) 64-Bit Server VM (build 25.40-b25, mixed mode)
```

(2) 安装 Xcode、Xcode command line tools 和 iOS 模拟器。

在 App Store 中下载 xcode.dmg 并安装。注意,要与 Mac 的 OS X 版本对应。由于 App 源代码是在 Xcode 开发环境中开发出来的,因此我们需要在 Xcode 里调试源码成功运行,并启动所要测试的 App。源码的导入和调试可以找 iOS 开发负责人帮忙,模拟器安装如图 3.1 所示。

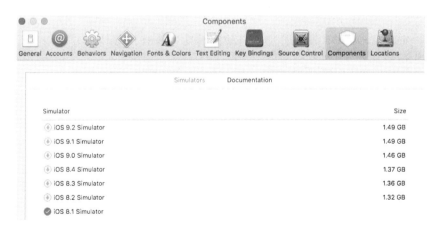

▲图 3.1

（3）安装 Homebrew：

curl -LsSf http://github.com/mxcl/homebrew/tarball/master | sudo tar xvz -C/usr/local --strip 1

brew –v 显示如下所示：

```
zouhuideMacBook-Air:~ zouhui$ brew -v
Homebrew 0.9.8 (git revision be7a; last commit 2015-09-15)
Homebrew/homebrew-core N/A
```

（4）安装 indeviceinstaller：

brew install indeviceinstaller

ideviceinstaller –h 显示如下所示：

```
zouhuideMacBook-Air:~ zouhui$ ideviceinstaller -h
Usage: ideviceinstaller OPTIONS
Manage apps on iOS devices.
```

（5）安装 Appium 服务端 Node.js：

brew install node

node–v 显示如下所示：

```
zouhuideMacBook-Air:~ zouhui$ node -v
v0.12.7
```

（6）安装 Appium 客户端。双击安装包 appium.dmg，或将它拖到 Applications 文件夹即完成安装。设置 Appium 环境变量，将 appium.js 和 appium-doctor.js 文件所在路径配置到 PATH 环境变量，如图 3.2 所示。

```
# Setting PATH for Python 2.7
# The orginal version is saved in .bash_profile.pysave
PATH="/Library/Frameworks/Python.framework/Versions/2.7/bin:${PATH}"
export PATH

export JAVA_HOME=$(/usr/libexec/java_home)

export NODE_HOME="/Applications/devtools/nodejs"
export PATH=${PATH}:${NODE_HOME}/bin

export PATH="/Applications/Appium.app/Contents/Resources/node/bin/":$PATH
export PATH="/Applications/Appium.app/Contents/Resources/node_modules/appium/bin/":$PATH
export ANDROID_HOME="/Users/young/Library/Android/sdk"
export PATH="/Users/zouhui/Library/Android/sdk/platform-tools":$PATH
export PATH="/Users/zouhui/Library/Android/sdk/tools":$PATH
export JAVA_HOME="/Library/Java/JavaVirtualMachines/jdk1.7.0_79.jdk/Contents/Home"
~
```

▲图 3.2

打开 Appium 客户端，如图 3.3 所示。

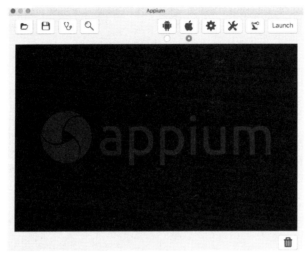

▲图 3.3

（7）用 npm 安装 wd：

```
npm install wd
```

（8）安装 SourceTree、Git 或 SVN，然后取开发的 App 源代码：

```
clonessh://zouhui@XXX.XXXXX.com:30020/mnt/repo/app
git 用户名密码：zouhui/3ihU88uy＊＊＊     版本管理取源码，输入用户名密码
ssh://zouh@dbs.e.com:30020/mnt/repo/ev＊＊_  App  版本管理取源码路径
```

（9）申请开发者调试证书。

因为开发者证书苹果是要收费的，所以需要请 iOS 团队开发负责人在后端平台工具申请自动化测试所需要的开发源码和调试权限，也可向公司的 iOS 开发负责人提出申请，提供手机 UDID: 3c22f4c14660eda7d3051636ae659b6b998af8db。

然后 iOS 开发负责人会提供：iOS 的 App 源代码、调试证书、开通手机 UDID 调试权限。

建议请 iOS 开发负责人直接安装好调试证书，因为证书安装比较麻烦，有时很容易出错，所以请开发人员安装可以很快解决。

也可以先在模拟器上运行，但一般真实测试过程中都是在真机上运行。

（10）安装 Eclipse。

（11）安装 Ant，并设置环境变量：

```
sudo
chmod +w /etc/bashrc
export ANT_HOME=/usr/local/apache-ant-1.9.3
export PATH=${PATH}:${ANT_HOME}/bin
```

（12）Appium 客户端界面的配置，真机测试时需要设置以下 4 个必填项。

1）BundleID：待测试 App 的 package 名称，如 com.test.buy。

2）Force Device：iPhone 手机设备，如 iPhone 6。

3）Platform Version：iOS 版本号，如 9.3。

4）UDID：手机设备唯一 ID，如 3c22f4c14660eda7d3051636ae659b6b998af7dc。

如图 3.4 所示。

▲图 3.4

Appium 中 Xcode 默认 Path 设置如图 3.5 所示。

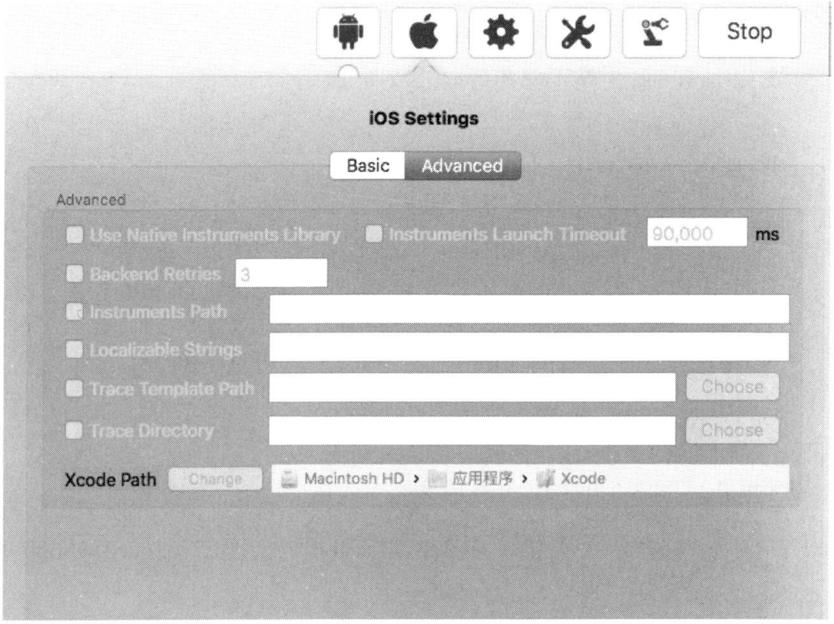

▲图 3.5

Appium 中的环境配置检查 Doctor，打勾表示正常，如图 3.6 所示。

▲图 3.6

Appium 中的 Inspector 工具，可以帮助我们快速定位元素，通常也是通过 name 和 XPath 等方式。

Appium 中，Launch 用来启动停止操作，启动成功时如图 3.8 所示。

▲图 3.7

单击图标 🔍，弹出 Inspector，进行控件元素的获取，查找元素的 XPath、name 等，如图 3.9 所示。

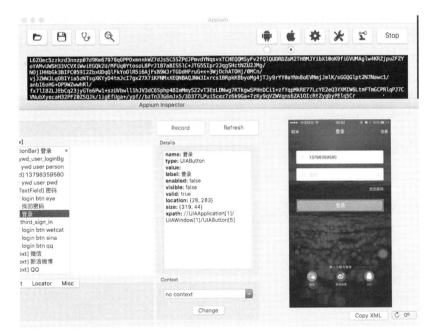

▲图 3.8

3.3　iOS 自动化测试运行

（1）iPhone 手机开启运行自动化，执行"设置→开发者→Enable UI Automation→yes"操作。

将手机连接到电脑上，并且使 Xcode 能识别到该手机。

（2）单击"Launch"按钮，启动 Appium，无错误日志。

（3）在 Eclipse 中运行已经写好的测试代码（请看第 4 章），可以看到，在 Appium 客户端中，有自动化数据和详细日志信息输出，手机上可以看到 App 自动化测试用例已开始运行。

（4）查看测试结果。

3.4　iOS 的 App 自动化测试 demo 演示视频

通常我们在做自动化测试框架编写的时候，前期可以给项目组测试人员，开发人员，团队 leaders 做 demo 演示和讲解，以增加大家对自动化测试的认识，信任，要求和期望。

登录账号：7980068@qq.com

登录密码：*******

http://www.iqiyi.com/w_19rsoh2gip.html#vfrm=2-3-0-1

http://v.youku.com/v_show/id_XMTY2MzU4OTk5Ng==.html

第 4 章

App 自动化测试源代码

4.1 基于 Java 的 App 自动化源代码解析

基于 Java 的 App 自动化源代码可同时支持 Android 和 iOS，仅需修改 Config.Java 中的配置文件即可。使用 Java 语言编写，代码简洁，功能强大，能够做到增改删用例时几乎不用写代码，只需在代码中简单配置一下用例编号信息即可，完整代码包如图 4.1 所示。

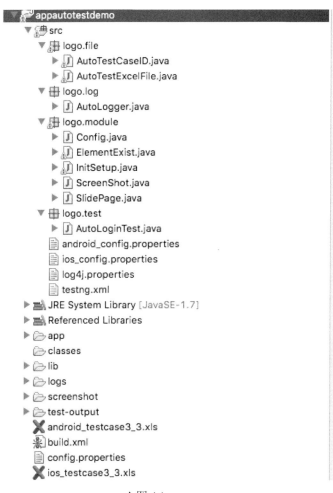

▲图 4.1

接下来，我们看看 App 测试自动化框架的代码。

程序清单 4-1　Java 代码

```java
logo.file
AutoTestCaseID.java

package logo.file;

import io.appium.java_client.AppiumDriver;
import io.appium.java_client.TouchAction;
import java.io.IOException;
import jxl.read.biff.BiffException;
import logo.module.ElementExist;
import logo.module.SlidePage;

import org.openqa.selenium.By;
import org.openqa.selenium.WebElement;

//读取一条测试用例的类
public class AutoTestCaseID
{
    ElementExist el = new ElementExist();                      // 创建判断元素是否存在的类的对象并初始化
    static AutoTestExcelFile ft = new AutoTestExcelFile();     // 创建读取 Excel 文件数据类的静态对象并初始化
    SlidePage sp = new SlidePage();                            // 创建页面滑动类的对象并初始化

    public void TestcaseId(AppiumDriver driver, String id) throws InterruptedException, BiffException,
IOException
    {                                       // 根据每一行循环，以及每一列循环，根据标题
                                            // 判断该行该列即固定单元格的值，并进行相应
                                            // 的操作
            int i, j, k, l, m, n, o, p, q;  // 定义变量，除变量 i 外每一个变量名代表一
                                            // 列标题的数值
        for (i = 0; i < ft.ReadContent().size(); i++)
        {
            if (ft.ReadContent().get(i).contains(id))
            {
                for (j = 0; j < ft.ReadTitle().size(); j++)
                {
```

```
            if (ft.ReadTitle().get(j).contains("定位方式"))
            {
                break;
            }
        }
        String caseidLocation = ft.ReadTitleContent(i + 1, j);

        for (k = 0; k < ft.ReadTitle().size(); k++)
        {
            if (ft.ReadTitle().get(k).contains("控件元素"))
            {
                break;
            }
        }
        String caseidElement = ft.ReadTitleContent(i + 1, k);

        for (l = 0; l < ft.ReadTitle().size(); l++)
        {
            if (ft.ReadTitle().get(l).contains("操作方法"))
            {
                break;
            }
        }
        String caseidOperationMethod = ft.ReadTitleContent(i + 1, l);

        for (m = 0; m < ft.ReadTitle().size(); m++)
        {
            if (ft.ReadTitle().get(m).contains("测试数据"))
            {
                break;
            }
        }
        String caseidTestData = ft.ReadTitleContent(i + 1, m);

        for (n = 0; n < ft.ReadTitle().size(); n++)
        {
            if (ft.ReadTitle().get(n).contains("验证数据"))
            {
                break;
            }
```

```
}
String caseidVerifyData = ft.ReadTitleContent(i + 1, n);

for (o = 0; o < ft.ReadTitle().size(); o++)
{
    if (ft.ReadTitle().get(o).contains("延迟时间"))
    {
        break;
    }
}
String SleepTime = ft.ReadTitleContent(i + 1, o);

if (caseidLocation.equals("By.xpath"))
{
    if (caseidOperationMethod.equals("sendkeys"))
    {
        el.waitForElementByXpath(caseidElement, driver);
        driver.findElement(By.xpath(caseidElement)).sendKeys(caseidTestData);
        if (SleepTime != null && SleepTime.length() != 0)
        {
            String StepTime = SleepTime.substring(0, SleepTime.indexOf("."));
            Thread.sleep(Integer.parseInt(StepTime));
        }
    } else if (caseidOperationMethod.equals("click"))
    {
        el.waitForElementByXpath(caseidElement, driver);
        driver.findElement(By.xpath(caseidElement)).click();
        if (SleepTime != null && SleepTime.length() != 0)
        {
            String StepTime = SleepTime.substring(0, SleepTime.indexOf("."));
            Thread.sleep(Integer.parseInt(StepTime));
        }
    } else if (caseidOperationMethod.equals("swipedown"))
    {
        el.waitForElementByXpath(caseidElement, driver);
        sp.Down_Page(caseidElement, driver);
        if (SleepTime != null && SleepTime.length() != 0)
        {
            String StepTime = SleepTime.substring(0, SleepTime.indexOf("."));
            Thread.sleep(Integer.parseInt(StepTime));
        }
```

```java
                }
        } else if (caseidLocation.equals("By.id"))
        {
            if (caseidOperationMethod.equals("sendkeys"))
            {
                el.waitForElementById(caseidElement, driver);
                driver.findElement(By.id(caseidElement)).sendKeys(caseidTestData);
                if (SleepTime != null && SleepTime.length() != 0)
                {
                    // String StepTime =
                    // SleepTime.substring(0,SleepTime.indexOf("."));
                    Thread.sleep(Integer.parseInt(SleepTime));
                }
            } else if (caseidOperationMethod.equals("click"))
            {
                el.waitForElementById(caseidElement, driver);
                driver.findElement(By.id(caseidElement)).click();
                if (SleepTime != null && SleepTime.length() != 0)
                {
                    // String StepTime =
                    // SleepTime.substring(0,SleepTime.indexOf("."));
                    Thread.sleep(Integer.parseInt(SleepTime));
                }
            } else if (caseidOperationMethod.equals("swipedown"))
            {
                el.waitForElementById(caseidElement, driver);
                sp.Down_Page(caseidElement, driver);
                if (SleepTime != null && SleepTime.length() != 0)
                {
                    String StepTime = SleepTime.substring(0, SleepTime.indexOf("."));
                    Thread.sleep(Integer.parseInt(StepTime));
                }
            } else if (caseidOperationMethod.equals("DownPage"))
            {
                el.waitForElementById(caseidElement, driver);
                sp.DownPage(caseidElement, driver);
                if (SleepTime != null && SleepTime.length() != 0)
                {
                    // String StepTime =
                    // SleepTime.substring(0,SleepTime.indexOf("."));
```

```java
                    Thread.sleep(Integer.parseInt(SleepTime));
                }
            }
        } else if (caseidLocation.equals("By.name"))
        {
            if (caseidOperationMethod.equals("sendkeys"))
            {
                el.waitForElementByName(caseidElement, driver);
                driver.findElement(By.name(caseidElement)).sendKeys(caseidTestData);
                if (SleepTime != null && SleepTime.length() != 0)
                {
                    String StepTime = SleepTime.substring(0, SleepTime.indexOf("."));
                    Thread.sleep(Integer.parseInt(StepTime));
                }
            } else if (caseidOperationMethod.equals("click"))
            {
                el.waitForElementByName(caseidElement, driver);
                driver.findElement(By.name(caseidElement)).click();
                if (SleepTime != null && SleepTime.length() != 0)
                {
                    // String StepTime =
                    // SleepTime.substring(0,SleepTime.indexOf("."));
                    Thread.sleep(Integer.parseInt(SleepTime));
                }
            } else if (caseidOperationMethod.equals("press"))
            {
                el.waitForElementByName(caseidElement, driver);
                WebElement e = driver.findElement(By.name(caseidElement)); // 长按
                                                                          // 拍视频
                TouchAction action = new TouchAction(driver);
                action.press(e).waitAction(5000);
                action.perform();
                if (SleepTime != null && SleepTime.length() != 0)
                {
                    String StepTime = SleepTime.substring(0, SleepTime.indexOf("."));
                    Thread.sleep(Integer.parseInt(StepTime));
                }
            }
        } else if (caseidLocation.equals("By.driver"))
        {
            if (caseidOperationMethod.equals("swipetoup"))
```

```java
                {
                    // el.waitForElementById(caseidObjectEntity, driver);
                    sp.swipeToUp(Integer.parseInt(caseidElement), driver);
                    if (SleepTime != null && SleepTime.length() != 0)
                    {
                        String StepTime = SleepTime.substring(0, SleepTime.indexOf("."));
                        Thread.sleep(Integer.parseInt(StepTime));
                    }
                }
        } else if (caseidLocation.equals("By.name"))
        {
            if ("pass".equals(el.waitForElementByNameSkip(caseidElement, driver)))
            {
                if (caseidOperationMethod.equals("click"))
                {
                    // el.waitForElementByName(caseidObjectEntity,
                    // driver);
                    driver.findElement(By.name(caseidElement)).click();
                    if (SleepTime != null && SleepTime.length() != 0)
                    {
                        String StepTime = SleepTime.substring(0, SleepTime.indexOf("."));
                        Thread.sleep(Integer.parseInt(StepTime));
                    }
                }
            } else if ("failed".equals(el.waitForElementByNameSkip(caseidElement, driver)))
            {
                continue;
            }
        }

        if (caseidVerifyData != null && caseidVerifyData.length() != 0)
        {
            el.waitForElementByName(caseidVerifyData, driver);
        }

        for (p = 0; p < ft.ReadTitle().size(); p++)
        {
            if (ft.ReadTitle().get(p).contains("测试结果"))
            {
                break;
            }
```

```
                        }
                        ft.WriteTitleContent(i + 1, (short) p);
                } else
                {

                }
            }
            Thread.sleep(3000);
        }
}
```

程序清单 4-2　Java 代码

logo.file
AutoTestExcelFile.java
package logo.file;

import java.io.FileInputStream;
import java.io.FileNotFoundException;
import java.io.FileOutputStream;
import java.io.IOException;
import java.io.InputStream;
import java.text.DecimalFormat;
import java.util.ArrayList;
import java.util.Date;
import java.util.HashMap;
import java.util.List;
import java.util.Map;

import jxl.read.biff.BiffException;
import logo.module.Config;
import logo.module.ElementExist;

import org.apache.poi.hssf.usermodel.HSSFCell;
import org.apache.poi.hssf.usermodel.HSSFRichTextString;
import org.apache.poi.hssf.usermodel.HSSFRow;
import org.apache.poi.hssf.usermodel.HSSFSheet;
import org.apache.poi.hssf.usermodel.HSSFWorkbook;
import org.apache.poi.poifs.filesystem.POIFSFileSystem;

//操作 Excel 表格中数据的类

```java
public class AutoTestExcelFile
{
    private POIFSFileSystem filesystem;
    private HSSFWorkbook workbook;
    private HSSFSheet sheet;
    private HSSFRow row;
    private HSSFCell cell;
    ElementExist el = new ElementExist();

    public String[] readExcelTitleContent(InputStream is)              // 读取 Excel 表格标题的内容
    {
        try
        {
            filesystem = new POIFSFileSystem(is);                       // 载入 Excel 文件
            workbook = new HSSFWorkbook(filesystem);
        } catch (IOException e)
        {
            e.printStackTrace();
        }
        sheet = workbook.getSheet(Config.getInstance().getCfg("My"));
// 读取 android_config.properties 或 ios_config.properties 配置文件中 My=我对应 Excel 中 "我" 的 Sheet 页
        row = sheet.getRow(0);                                          // 默认第一列
        int colNum = row.getPhysicalNumberOfCells();                    // 标题总列数
        String[] title = new String[colNum];
        for (int i = 0; i < colNum; i++)
        {
            title[i] = getStringCellValue(row.getCell((short) i));      // 标题的内容
        }
        return title;                                                   // 返回标题内容数组数据
    }

    public Map<Integer, String> readExcelContent(InputStream is)        // 读取 Excel 数据内容
    {
        Map<Integer, String> content = new HashMap<Integer, String>();
        String str = "";
        try
        {
            filesystem = new POIFSFileSystem(is);
            workbook = new HSSFWorkbook(filesystem);
        } catch (IOException e)
```

```
        {
            e.printStackTrace();
        }
        sheet = workbook.getSheet(Config.getInstance().getCfg("My"));  // 得到 Excel 为"我"的 Sheet 表格
        int rowNum = sheet.getLastRowNum();                            // 得到总行数
        row = sheet.getRow(0);
        int colNum = row.getPhysicalNumberOfCells();
        for (int i = 1; i <= rowNum; i++)            // 正文内容应该从第二行开始,第一行为表头的标题
        {
            row = sheet.getRow(i);
            int j = 0;
            while (j < colNum)
            {                                        // 每个单元格的数据内容用"-"分隔开,以后需要时
                                                     // 用 String 类的 replace()方法还原数据
                str += getStringCellValue(row.getCell((short) j)).trim() + "-";
                                                     // 也可以将每个单元格的数据设置到一个 Javabean 的
                                                     // 属性中,此时需要新建一个 Javabean \
                j++;
            }
            content.put(i, str);
            str = "";
        }
        return content;
    }

    private String getStringCellValue(HSSFCell cell)    // 获取 Excel 单元格数据内容中为字符串类型的数据
    {
        String strCell = "";
        if (cell == null)
        {
            return "";
        } else
            switch (cell.getCellType())
            {
            case HSSFCell.CELL_TYPE_STRING:
                strCell = cell.getStringCellValue();
                break;
            case HSSFCell.CELL_TYPE_NUMERIC:
                String strCellNumber = String.valueOf(cell.getNumericCellValue());
                DecimalFormat df = new DecimalFormat("0");
```

```
                    strCell = df.format(cell.getNumericCellValue());
                    break;
            case HSSFCell.CELL_TYPE_BOOLEAN:
                    strCell = String.valueOf(cell.getBooleanCellValue());
                    break;
            case HSSFCell.CELL_TYPE_BLANK:
                    strCell = "";
                    break;
            default:
                    strCell = "";
                    break;
        }
        if (strCell.equals("") || strCell == null)
        {
            return "";
        }

        return strCell;
}

private String getDateCellValue(HSSFCell cell)          //获取 Excel 单元格数据内容中为日期类型的数据
{
    String result = "";
    try
    {
        int cellType = cell.getCellType();
        if (cellType == HSSFCell.CELL_TYPE_NUMERIC)
        {
            Date date = cell.getDateCellValue();
            result = (date.getYear() + 1900) + "-" + (date.getMonth() + 1) + "-" + date.getDate();
        } else if (cellType == HSSFCell.CELL_TYPE_STRING)
        {
            String date = getStringCellValue(cell);
            result = date.replaceAll("[年月]", "-").replace("日", "").trim();
        } else if (cellType == HSSFCell.CELL_TYPE_BLANK)
        {
            result = "";
        }
    } catch (Exception e)
    {
```

```java
            System.out.println("日期格式不正确!");
            e.printStackTrace();
        }
        return result;
    }

    public List<String> ReadTitle() throws FileNotFoundException        //读取 Excel 标题内容
    {
        List<String> list = new ArrayList<String>();                    // 对读取 Excel 表格标题测试
        InputStream is = new FileInputStream(Config.getInstance().getCfg("filePath"));
        AutoTestExcelFile excelReader = new AutoTestExcelFile();
        String[] title = excelReader.readExcelTitleContent(is);
        // System.out.println("获得 Excel 表格的标题:");
        for (String s : title)
        {
            // System.out.print(s + " ");
            list.add(s);
        }
        return list;
    }

    public List<String> ReadContent() throws FileNotFoundException      //读取 Excel 表格中测试用例的内容
    {
        List<String> list = new ArrayList<String>();
        AutoTestExcelFile excelReader = new AutoTestExcelFile();
        InputStream is2 = new FileInputStream(Config.getInstance().getCfg("filePath"));
        Map<Integer, String> map = excelReader.readExcelContent(is2);
        for (int i = 1; i <= map.size(); i++)
        {
            list.add(map.get(i));
        }
        return list;
    }

    public String ReadTitleContent(int i, int j) throws BiffException, IOException
    {
        InputStream is = new FileInputStream(Config.getInstance().getCfg("filePath"));
        try
        {
            filesystem = new POIFSFileSystem(is);
```

```java
            workbook = new HSSFWorkbook(filesystem);
        } catch (IOException e)
        {
            e.printStackTrace();
        }
        sheet = workbook.getSheet(Config.getInstance().getCfg("My"));
        row = sheet.getRow(i);
        String content = getStringCellValue(row.getCell((short) j));
        return content;
    }

    private void saveWorkBook(HSSFWorkbook wb)
    {
        try
        {
            FileOutputStream fileOut = new FileOutputStream(Config.getInstance().getCfg("filePath"));
            wb.write(fileOut);
        } catch (FileNotFoundException ex)
        {
            System.out.println(ex.getMessage());
        } catch (IOException ex)
        {
            System.out.println(ex.getMessage());
        }
    }

    private HSSFCell getCell(HSSFSheet sheet, int rowIndex, short columnIndex)
    {
        HSSFRow row = sheet.getRow(rowIndex);
        if (row == null)
        {
            row = sheet.createRow(rowIndex);
        }
        HSSFCell cell = row.getCell(columnIndex);
        if (cell == null)
        {
            cell = row.createCell((short) columnIndex);
        }
        return cell;
    }
```

```java
public void WriteTitleContent(int i, short j) throws BiffException, IOException    //记录测试结果，当元素
                                                                                   //存在时为 Pass，否则为空
{
    try
    {
        InputStream is = new FileInputStream(Config.getInstance().getCfg("filePath"));
        filesystem = new POIFSFileSystem(is);
        workbook = new HSSFWorkbook(filesystem);
        is.close();
    } catch (IOException ex)
    {
        System.out.println(ex.getMessage());
    }

    sheet = workbook.getSheet(Config.getInstance().getCfg("My"));
    row = sheet.getRow(i);
    HSSFCell cell = getCell(sheet, i, j);
    HSSFRichTextString hts = new HSSFRichTextString(el.result);
    cell.setCellValue(hts);
    saveWorkBook(workbook);
}

public void SetContentInit(short j) throws BiffException, IOException     //执行测试用例前，初始化设置固
//定列的数据为空；即默认测试结果为空，也就是在执行测试前，将测试结果数据置为空
{
    try
    {
        InputStream is = new FileInputStream(Config.getInstance().getCfg("filePath"));
        Map<Integer, String> map = readExcelContent(is);

        for (int k = 1; k <= map.size(); k++)
        {
            sheet = workbook.getSheet(Config.getInstance().getCfg("My"));
            row = sheet.getRow(k);
            HSSFCell cell = getCell(sheet, k, j);
            HSSFRichTextString hts = new HSSFRichTextString("");
            cell.setCellValue(hts);
        }
        saveWorkBook(workbook);
```

```
            } catch (IOException ex)
            {
                    System.out.println(ex.getMessage());
            }
        }
}
```

程序清单 4-3　　Java 代码

logo.log
AutoLogger.java
package logo.log;

import java.io.FileInputStream;
import java.io.IOException;
import java.io.InputStream;
import java.text.SimpleDateFormat;
import java.util.Calendar;
import java.util.Properties;

import org.apache.log4j.Logger;
import org.apache.log4j.PropertyConfigurator;
import org.testng.Reporter;

```java
public class AutoLogger
{
        private static Logger logger = null;
        private static AutoLogger logg = null;

        public static AutoLogger getLogger(Class<?> T)
        {
            if (logger == null)
            {
                Properties props = new Properties();
                try
                {
                        InputStream is = new FileInputStream("src//log4j.properties");
                        props.load(is);
                } catch (IOException e)
                {
                        e.printStackTrace();
```

```
                }
                PropertyConfigurator.configure(props);
                logger = Logger.getLogger(T);
                logg = new AutoLogger();
            }
        return logg;
    }

    // 重写 logger 方法
    public void log(String msg)
    {
        SimpleDateFormat sdf = new SimpleDateFormat("yyyy-MM-dd HH:mm:ss");
        Calendar ca = Calendar.getInstance();
        logger.info(msg);
        Reporter.log("Reporter:" + sdf.format(ca.getTime()) + "===>" + msg);
    }

    public void debug(String msg)
    {
        logger.debug(msg);
    }

    public void warn(String msg)
    {
        logger.warn(msg);
        Reporter.log("Reporter:" + msg);
    }

    public void error(String msg)
    {
        logger.error(msg);
        Reporter.log("Reporter:" + msg);
    }
}
```

程序清单 4-4　Java 代码

logo.moidue
Config.java
package logo.module;

```java
import java.io.IOException;
import java.io.InputStream;
import java.util.HashMap;
import java.util.Map;
import java.util.Map.Entry;
import java.util.Properties;

public class Config
// 读取配置文件信息的类，当 CONFIG_FILE＝"android_config.properties"将读取 android 配置文件的信息
{
    static final String CONFIG_FILE = "android_config.properties";
// 当 CONFIG_FILE＝"ios_config.properties"将读取 ios 配置文件的信息
    private Map<String, String> configMap = new HashMap<String, String>();
    private static Config testResManager;

    public static Config getInstance()
    {
        if (testResManager == null)
        {
            testResManager = new Config();
        }
        return testResManager;
    }

    public Config()
    {
        configMap = loadFile(Config.CONFIG_FILE);
    }

    protected Map<String, String> loadFile(String fileName)
    {
        Map<String, String> map = null;
        if ((fileName != null) && (!fileName.trim().equals("")))
        {
            InputStream is = Config.class.getClassLoader().getResourceAsStream(fileName);
            if (is != null)
            {
                map = new HashMap<String, String>();
                Properties prop = new Properties();
                try
```

```java
            {
                    prop.load(is);
                    for (Entry<Object, Object> entry : prop.entrySet())
                    {
                            String key = (String) entry.getKey();
                            String value = (String) entry.getValue();
                            map.put(key, value);
                    }
            } catch (IOException e)
            {
                    e.printStackTrace();
            }
        }
    }
    return map;
}

public String getCfg(String key)
{
        return configMap.get(key);
}

}
```

程序清单 4-5　Java 代码

logo.module
ElementExist.java

```java
package logo.module;

import io.appium.java_client.AppiumDriver;
import org.openqa.selenium.By;
import org.openqa.selenium.support.ui.ExpectedConditions;
import org.openqa.selenium.support.ui.WebDriverWait;
import org.testng.Assert;

public class ElementExist // 判断元素是否存在的类，当元素不存在或错误时，会自动截屏保存错误信息的图片
{
        public static String result;
```

```java
public String waitForElementByXpath(final String ID, AppiumDriver driver)
{
    try
    {
        WebDriverWait wait = new WebDriverWait(driver, 18);
// 设置智能等待和超时时间，即 18 秒内，判断到元素存在则立即跳出等待，直接操作它
        wait.until(ExpectedConditions.presenceOfElementLocated(By.xpath(ID)));
        if (null == wait)
        {
            ScreenShot.takeScreenShot(driver);
// 根据元素 Xpath 判断，若超过等待时间，则自动截屏保存错误图片
            result = "failed";
            Assert.assertTrue(false);
        } else
        {
            result = "pass";
        }
    } catch (Exception e)
    {
        ScreenShot.takeScreenShot(driver);
// 根据元素 Xpath 判断,若元素不存在，则自动截屏保存错误图片
        result = "failed";
        Assert.assertTrue(false);
    }
    return result;
}

public String waitForElementByName(final String ID, AppiumDriver driver)
{
    try
    {
        WebDriverWait wait = new WebDriverWait(driver, 18);
        wait.until(ExpectedConditions.presenceOfElementLocated(By.name(ID)));
        if (null == wait)
        {
            ScreenShot.takeScreenShot(driver);
// 根据元素名称判断，若超过等待时间，则自动截屏保存错误图片
            result = "failed";
            Assert.assertTrue(false);
        } else
```

```
            {
                result = "pass";
            }
        } catch (Exception e)
        {
            ScreenShot.takeScreenShot(driver);
// 根据元素名称判断,当元素不存在时,自动截屏保存错误图片
            result = "failed";
            System.out.print(result);
            Assert.assertTrue(false);
        }
        return result;
    }

    public String waitForElementByNameSkip(final String ID, AppiumDriver driver)
    {
        try
        {
            WebDriverWait wait = new WebDriverWait(driver, 18);
            wait.until(ExpectedConditions.presenceOfElementLocated(By.name(ID)));
            if (null == wait)
            {
                ScreenShot.takeScreenShot(driver);
                result = "failed";
            } else
            {
                result = "pass";
            }
        } catch (Exception e)
        {
            ScreenShot.takeScreenShot(driver);
            result = "failed";
            Assert.assertTrue(false);
        }
        return result;
    }

    public String waitForElementById(final String ID, AppiumDriver driver)
    {
        try
        {
```

```java
            WebDriverWait wait = new WebDriverWait(driver, 18);
            wait.until(ExpectedConditions.presenceOfElementLocated(By.id(ID)));
            if (null == wait)
            {
                    ScreenShot.takeScreenShot(driver);
// 根据元素 id 判断，若超过等待时间，则自动截屏保存错误图片
                    result = "failed";
                    Assert.assertTrue(false);
            } else
            {
                    result = "pass";
            }

        } catch (Exception e)
        {
                ScreenShot.takeScreenShot(driver);
// 根据元素 id 判断,若元素不存在，则自动截屏保存错误图片

                result = "failed";
                Assert.assertTrue(false);
        }
        return result;
    }
}
```

程序清单 4-6　Java 代码

logo.module
InitSetup.java
package logo.module;

import io.appium.java_client.AppiumDriver;

import java.io.File;
import Java.net.MalformedURLException;
import org.openqa.selenium.remote.CapabilityType;
import org.openqa.selenium.remote.DesiredCapabilities;

public class InitSetup
// 初始化配置文件信息的类，并自动根据配置文件是 android_config.properties 还是 ios_config.properties
// 进行判断

```java
{
    public DesiredCapabilities InitSetupCFG(final DesiredCapabilities capabilities) throws MalformedURLException
    { // set up appium
        if (Config.CONFIG_FILE.equals("android_config.properties"))
        //对配置文件进行判断,如果是 android_config.properties,则读取 Android 配置信息,默认为 Android
        {
            final File appDir = new File(System.getProperty("user.dir"), "app");
            final File app = new File(appDir, Config.getInstance().getCfg("appDir"));
            capabilities.setCapability("app", Config.getInstance().getCfg("app"));
            //设置 App 所在路径
            capabilities.setCapability(CapabilityType.BROWSER_NAME, "");
            capabilities.setCapability("platformVersion", Config.getInstance().getCfg("platformVersion"));
            //设置 Android 版本号
            capabilities.setCapability("platformName", Config.getInstance().getCfg("platformName"));
            //设置 Android 系统
            capabilities.setCapability("deviceName", Config.getInstance().getCfg("deviceName"));
            //设置所连接的 Android 手机名称
            capabilities.setCapability("appPackage", Config.getInstance().getCfg("AppPackage"));
            //设置 App 包名 Package
            capabilities.setCapability("appActivity", Config.getInstance().getCfg("AppActivity"));
            //设置 App 启动入口 Activity
            capabilities.setCapability("unicodeKeyboard", Config.getInstance().getCfg("unicodeKeyboard"));
            //设置支持中文输入
            capabilities.setCapability("resetKeyboard", Config.getInstance().getCfg("resetKeyboard"));
            //重置 Appium 为默认输入法
            capabilities.setCapability("newCommandTimeout", Config.getInstance().getCfg("newCommandTimeout"));
            //设置 Appium 命令超时时间
        } else if (Config.CONFIG_FILE.equals("ios_config.properties"))
        //对配置文件进行判断,如果是 ios_config.properties,则读取 iOS 配置信息,则为运行苹果手机 App 的测试
        //用例
        {
            final File AppDir = new File(System.getProperty("user.dir"), "App");
            final File App = new File(AppDir, Config.getInstance().getCfg("appDir"));
            //设置 App 所在路径
            capabilities.setCapability("platformVersion", Config.getInstance().getCfg("platformVersion"));
            //设置 iOS 版本号
            capabilities.setCapability("platformName", Config.getInstance().getCfg("platformName"));
            //设置 iOS 系统
            capabilities.setCapability("deviceName", Config.getInstance().getCfg("deviceName"));
```

```
//设置所连接的 iPhone 手机名称
                capabilities.setCapability("udid", Config.getInstance().getCfg("udid"));
//设置 iPhone 手机的 UDID
                capabilities.setCapability("app", Config.getInstance().getCfg("app"));
//设置 iOS 启动 App 包名
                return capabilities;
        }
        return capabilities;
    }

    public void TearDownCFG(AppiumDriver driver) throws MalformedURLException
    {
        driver.quit();
    }
}
```

程序清单 4-7 Java 代码

```
logo.module
ScreenShot.java
package logo.module;

import java.io.File;
import java.io.IOException;
import java.text.SimpleDateFormat;
import java.util.Date;

import org.apache.commons.io.FileUtils;
import org.openqa.selenium.OutputType;
import org.openqa.selenium.TakesScreenshot;
import org.openqa.selenium.WebDriver;

public class ScreenShot
// 截屏保存图片的类,并自动根据当天日期分天保存在这个文件夹下面
{
    public static void takeScreenShot(WebDriver driver)
    {
        File screenShotFile = ((TakesScreenshot) driver).getScreenshotAs(OutputType.FILE);
//截屏图片保存到路径以及取名
        try
        {
```

```java
            FileUtils.copyFile(screenShotFile,
                    new File("screenshot/" + getDatePath() + "/" + getCurrentDateTime() + ".jpg"));
//以图片当前时间为名称保存到当前日期的目录下
        } catch (IOException e)
        {
            e.printStackTrace();
//I/O 异常处理
        }
    }

    public static String getCurrentDateTime()
//得到当前时间
    {
        SimpleDateFormat df = new SimpleDateFormat("yyyyMMdd_HHmmss");
//格式化日期，精确到秒
        return df.format(new Date());
    }

    public static File getDatePath()
//得到当前日期，格式的路径
    {
        Date date = new Date();
        String path = new SimpleDateFormat("yyyyMMdd").format(date);
//格式化日期精确到天
        File f = new File(path);
//创建以当前日期为名称的文件夹
        return f;
    }
}
```

程序清单 4-8　Java 代码

logo.module
SlidePage.java

package logo.module;

import java.util.HashMap;

import io.Appium.java_client.AppiumDriver;

```java
import org.openqa.selenium.By;
import org.openqa.selenium.Dimension;
import org.openqa.selenium.javascriptExecutor;
import org.openqa.selenium.Point;
import org.openqa.selenium.WebElement;
import org.openqa.selenium.remote.RemoteWebElement;

public class SlidePage
// 滑动操作类，引导页滑动，默认滑动3页；根据元素id向下滑动；根据屏幕上下左右滑动
{
    public void Guide_Page(AppiumDriver driver) throws InterruptedException
        {   //引导页滑动，默认滑动3页；根据元素id向下滑动
            for (int i = 0; i < 3; i++)
            {
                int startx = -1;
                int starty = -1;
                int endx = -1;
                int endy = -1;
                Thread.sleep(3000);
                WebElement screen = driver.findElementById("com.life:id/pager");
                Point point = screen.getLocation();
                Dimension size = screen.getSize();
                endx = point.getX();
                endy = point.getY() + size.getHeight() / 2;
                startx = point.getX() + size.getWidth() - 5;
                starty = endy;
                driver.swipe(startx, starty, endx, endy, 700);
            }
            Thread.sleep(3000);
            driver.findElement(By.id("com.life:id/next")).click();
            Thread.sleep(2000);
        }

    public void Down_Page(String xpath, AppiumDriver driver) throws InterruptedException
        {   //根据屏幕向下滑动
            JavascriptExecutor js = (JavascriptExecutor) driver;
            WebElement element = driver.findElementByXPath(xpath);
            HashMap<String, Double> flickObject = new HashMap<String, Double>();
            flickObject.put("startX", 0.5);
            flickObject.put("startY", 0.8);
```

```java
        flickObject.put("endX", 0.6);
        flickObject.put("endY", 0.2);
        flickObject.put("element", Double.valueOf(((RemoteWebElement) element).getId()));
        js.executeScript("mobile: flick", flickObject);
}

public void DownPage(String id, AppiumDriver driver) throws InterruptedException
{       //根据元素 id 向下滑动
        int starty = -1;
        int startx = -1;
        int endx = -1;
        int endy = -1;
        Thread.sleep(3000);
        WebElement screen = driver.findElementById(id);
        Point point = screen.getLocation();
        Dimension size = screen.getSize();
        endy = point.getY();
        endx = point.getX() + size.getHeight() / 2;
        starty = point.getY() + size.getWidth() - 5;
        startx = endx;
        driver.swipe(startx, starty, endx, endy, 700);
        Thread.sleep(3000);
}

public void Left_Page(String xpath, AppiumDriver driver) throws InterruptedException
{       //根据 xpath 向左滑动
        JavascriptExecutor js = (JavascriptExecutor) driver;
        WebElement element = driver.findElementByXPath(xpath);
        HashMap<String, Double> flickObject = new HashMap<String, Double>();
        flickObject.put("startX", 0.8);
        flickObject.put("startY", 0.5);
        flickObject.put("endX", 0.2);
        flickObject.put("endY", 0.5);
        flickObject.put("element", Double.valueOf(((RemoteWebElement) element).getId()));
        js.executeScript("mobile: flick", flickObject);
}

public void swipeToUp(int during, AppiumDriver driver)
{       //根据屏幕向上滑动
        int width = driver.manage().window().getSize().width;
```

```java
            int height = driver.manage().window().getSize().height;
            driver.swipe(width / 2, height * 3 / 4, width / 2, height / 4, during);
    }

    public void swipeToDown(int during, AppiumDriver driver)
    {    //根据屏幕向下滑动
            int width = driver.manage().window().getSize().width;
            int height = driver.manage().window().getSize().height;
            driver.swipe(width / 2, height / 4, width / 2, height * 3 / 4, during);
    }

    public void swipeToLeft(int during, AppiumDriver driver)
    {    //根据屏幕向左滑动
            int width = driver.manage().window().getSize().width;
            int height = driver.manage().window().getSize().height;
            driver.swipe(width * 3 / 4, height / 2, width / 4, height / 2, during);
    }

    public void swipeToRight(int during, AppiumDriver driver)
    {    //根据屏幕向右滑动
            int width = driver.manage().window().getSize().width;
            int height = driver.manage().window().getSize().height;
            driver.swipe(width / 4, height / 2, width * 3 / 4, height / 2, during);
    }

}
```

程序清单 4-9　Java 代码

```
logo.test
AutoLoginTest.java
package logo.test;

import io.Appium.java_client.AppiumDriver;
import java.io.IOException;
import java.net.MalformedURLException;
import java.net.URL;
import jxl.read.biff.BiffException;
import logo.file.AutoTestExcelFile;
import logo.file.AutoTestCaseID;
import logo.module.InitSetup;
```

```java
import org.openqa.selenium.remote.DesiredCapabilities;
import org.testng.annotations.BeforeClass;
import org.testng.annotations.Test;
import org.testng.annotations.BeforeMethod;
import org.testng.annotations.AfterMethod;

public class AutoLoginTest
{
    AppiumDriver driver;                                    // 创建 Appium 驱动对象
    AutoTestExcelFile fesm = new AutoTestExcelFile();       // 创建读取 Excel 文件类的对象并初始化
    InitSetup is = new InitSetup();                         // 创建初始配置文件信息类的对象并初始化
    AutoTestCaseID tcId = new AutoTestCaseID();             // 创建读取一条测试用例类的对象并初始化

    @BeforeClass
    public void beforeClass() throws BiffException, IOException    // 执行测试用例前
    {
        fesm.SetContentInit((short) 10);                    // 初始化 Excel 设置,"测试结果"默认为空
    }

    @BeforeMethod
    public void beforeMethod() throws MalformedURLException, InterruptedException
                                                            // 执行测试前
    {
        driver = new AppiumDriver(new URL("http://127.0.0.1:4723/wd/hub"), is.InitSetupCFG(new DesiredCapabilities()));
                                                            // 初始化 Appium 配置手机信息
    }

    @Test
    public void My_Login_001() throws InterruptedException, BiffException, IOException
                                                            // 执行登录的测试用例
    {
        tcId.TestcaseId(driver, "My_Login_001");            // 对应 Excel 里的测试用例一列,要完全一致
    }

    @AfterMethod
    public void afterMethod()                               // 执行测试后
    {
```

```
            driver.quit();                              // 退出 Appium 的当前 session 连接环境
    }
}
```

程序清单 4-10 （资源配置文件）

android_config.properties

------安卓手机、App、Appium 的相关连的配置信息------
appDir = F:/Users/Administrator/workspace/Apptest/App/20160704
platformVersion = 4.4   // Android 手机 Api 版本号
platformName = Android   // Android 手机操作系统
deviceName= C8817D   // 所连接的手机名称或 id
appPackage= com.seller2   // App 的 Package 包名
appActivity=start.StartActivity   // App 的 Activity 启动入口
unicodeKeyboard = True   // 设置 Appium 可以输入中文
resetKeyboard = True   // 设置 Appium 为默认输入法
newCommandTimeout = 300   // 设置命令超时时间为 300 秒

filePath = android_shop_testcase1_3.xls   // 读取测试用例的路径和文件
My = \u6211   // 读取 Excel 中 Sheet 表格的名称

ios_config.properties   资源配置文件

appDir =
/Users/zouhui/Library/Developer/CoreSimulator/Devices/11CDAEDE-291E-4B01-81BB-7E2C8C0BA65E/data/Containers/Bundle/Application/8BECE70C-F3E8-4AD3-B6C8-932B94239936/.app
platformVersion = 9.0
platformName = iOS
deviceName= iPhone 6
udid = 3c22f4c14660eda7d3051636ae659b6b998af8db
app = com.hele.ywbuyer

filePath = ios_testcase3_3.xls

程序清单 4-11   Java 代码（日志配置文件）

log4j.properties

log4j.rootLogger=Info, stdout, logfile

log4j.appender.stdout=org.apache.log4j.ConsoleAppender
log4j.appender.stdout.layout=org.apache.log4j.PatternLayout
log4j.appender.stdout.layout.ConversionPattern=%-d{yyyy-MM-dd HH:mm:ss,SSS} %p %t [%c]%M(line:%L)%m%n

log4j.appender.logfile.encoding=UTF-8
log4j.appender.logfile=org.apache.log4j.DailyRollingFileAppender
log4j.appender.logfile.File=logs/run.log
log4j.appender.logfile.layout=org.apache.log4j.PatternLayout
log4j.appender.logfile.layout.ConversionPattern=%-d{yyyy-MM-dd HH:mm:ss,SSS} %p %t %M(line:%L)%m%n

程序清单 4-12　　Java 代码（TestNG 配置文件）

testng.xml

```xml
<?xml version="1.0" encoding="UTF-8"?>
<suite name="Suite" parallel="false">
  <test name="Test_My_Login">
    <classes>
        <class name="logo.test.AutoLoginTest"/>
            <methods preserver-order="true">
                <include name="My_Login _001" />
            </methods>
    </classes>
  </test>

</suite> <!-- Suite -->
```

android_shop_testcase1.3.xls

（Android 测试用例 Excel 文件）

A	B	C	D	E	F	G	H	I	J	K
测试用例编号	用例描述	测试步骤	测试对象名称描述	定位方式	控件元素		操作方法	测试数据	验证数据	延迟时间	测试结果
My_Login_001	正常登录	TS-001	我的一登录页	By.name	登录	click				pass
My_Login_001	正常登录	TS-002	我的一密码文本框	By.id	com.test.seller:id/phone_ed	sendkeys	13798359580			pass
My_Login_001	正常登录	TS-005	我的一用户名文本框	By.id	com.test.seller:id/password	sendkeys	test123456			pass
My_Login_001	正常登录	TS-006	我的一登录按钮	By.id	com.test.seller:id/login_bt	click				pass
My_Login_001	正常登录	TS-007	我的一返回	By.name	我	click			10000	pass
My_Login_001	正常登录	TS-001	我的一首页	By.name	首页	click				pass
My_Login_001	正常登录	TS-002	我的一商品管理	By.name	商品管理	click				pass
My_Login_001	正常登录	TS-003	我的一历史商品	By.name	历史商品	click				pass
My_Login_001	正常登录	TS-004	我的一分类	By.name	分类	click				pass
My_Login_001	正常登录	TS-005	我的一销售中	By.name	销售中	click			8000	pass
My_Login_001	正常登录	TS-006	我的一向下滑动	By.id	com.test.seller:id/layout	DownPage			8000	pass

▲图 4.2

其中，My_Login_001 对应 MyLoginAutoTest 类中的方法名称及用例名称。

By.id 和 By.name 是 Android 对应的定位方式，可用来寻找控件元素，也可以是 XPath 等。

控件元素是通过 uiautoviewer 或 Inspector 工具捕获到的。

操作方法如下：click，表示单击某控件元素；sendkeys，表示在某控件元素中输入内容。

ios_testcase3.3.xls（iOS 测试用例 Excel 文件）

A	B	C	D	E	F	G	H	I	J	K
测试用例编号	用例描述	测试步骤	测试对象	定位方式	控件元素	操作方法	测试数据	验证数据	延迟时间	测试结果
My_Login_001	正常登录	TS-001	我的一首页	By.name	登录	click				pass
My_Login_001	正常登录	TS-002	我的一密码文本框	By.xpath	//UIAApplication[1]/UIAWindow[1]/UIASecureTextField[1]	sendkeys	test123456			pass
My_Login_001	正常登录	TS-003	我的一登录按钮	By.name	完成	click				pass
My_Login_001	正常登录	TS-004	我的一登录按钮	By.name	我	click				pass

▲图 4.3

这里只写了一个登录用例的自动化测试 demo，第二个以及批量自动化测试用例可以仿照这个用例来编写。大致如下，在 Excel 中写好第二个用例编号 ID（此编号最好能代表这个用例含义，第二个用例的各测试步骤（一个测试步骤为完整的一行，不要有空格或空行），控件的元素（通过 uiautomatorviewer 或 Inspector 等工具获取）、定位方式、操作方法、测试数据以及验证数据等都先一一的写好、写正确。然后在 AutoLoginTest.java 代码文件中加入以下代码（也可新建 TestNG 类文件）：

```
public void My_Login_001() throws InterruptedException, BiffException, IOException
                            // 执行登录的测试用例
{
tcId.TestcaseId(driver, "My_Login_001");    // 对应 Excel 里的编号一列
}
```

增加上述内容后，把 My_Login_001()改成第二个用例的编号 ID，即完成了第二个自动化测试用例的编写。

在 testng.xml 文件中加入一行<include name=" " />，并在引号中加入写好的要执行的第二个自动化测试用例编号 ID。增加批量用例时与此类似，读者可参考这个登录 demo 的用例去编写。

<methods preserver-order="true">

```xml
            <include name="My_Login_001" />
</methods>
```

由此可见，增、改、删自动化用例时，只需相应地修改 Excel 的用例编号、步骤等，以及修改 testNG 类中的用例编号和 testng.xml 文件中的运行用例编号。维护可以说是非常简洁，框架很强大，不需要增改代码逻辑或增改很多代码，即可快速应用于工作中，最大化地减少代码的维护成本和时间成本。

4.2　源代码结合 Ant 持续集成到 Jenkins

结合 TestNG 和 Ant build.xml（Ant 持续集成配置文件）：

```xml
<?xml version="1.0" encoding="UTF-8" ?>                 <!-- 设定中文编码 -->
<project name="iosTest" basedir="." default="run_tests">  <!-- 默认调用 run_tests 任务 -->
    <property name="src" value="src" />
    <property name="dest" value="classes" />
    <property name="lib.dir" value="${basedir}/lib" />
    <property name="output.dir" value="${basedir}/test-output" />  <!-- 设置报告输出的路径 -->

    <path id="compile.path">                            <!-- 编译路径设置 -->
        <fileset dir="${lib.dir}/">
            <include name="*.jar" />
        </fileset>
        <pathelement location="${src}" />
        <pathelement location="${dest}" />
    </path>

    <target name="init">                                <!-- 初始化设置 -->
        <mkdir dir="${dest}" />
    </target>

    <target name="compile" depends="init">              <!-- 编译和初始化 -->
        <echo>compile tests</echo>
        <javac srcdir="${src}" destdir="${dest}" encoding="UTF-8"
            classpathref="compile.path" />
    </target>
    <taskdef resource="testngtasks" classpath="${lib.dir}/testng.jar" />  <!--testing.jar 目录和文件 -->
```

```xml
<target name="run_tests" depends="compile">                    <!-- 开始测试 -->
    <echo>running tests</echo>
    <testng classpathref="compile.path" outputdir="${output.dir}"
        haltonfailure="no"
        failureproperty="failed"
        parallel="true"
        threadCount="1"  listeners="com.netease.qa.testng.PowerEmailableReporter,
        com.netease.qa.testng.RetryListener, com.netease.qa.testng.TestResultListener">
<!-- 监听器，调用如果失败则自动重复执行 -->
        <xmlfileset dir="${src}/">
            <include name="testng.xml" />
<!--结合 TestNG,调用 testng.xml 里面配置的测试用例  -->
        </xmlfileset>
        <classfileset dir="${dest}">
            <include name="/*.class" />
        </classfileset>

    </testng>
    <antcall target="transform" />
    <!-- <fail message="TEST FAILURE" if="failed" /> -->
</target>

<target name="transform" description="report">                 <!-- 生成报告 -->
    <xslt
        in="${output.dir}/testng-results.xml"
        style="${lib.dir}/testng-results.xsl"                  <!-- 以 testng-results.xsl 模板的方式 -->
        out="${output.dir}/Report.html"                        <!-- 输出 HTML 格式的测试报告 -->
        force="yes">
        <!-- you need to specify the directory here again -->
        <param name="testngXslt.outputDir" expression="${output.dir}" />
        <classpath refid="compile.path" />
    </xslt>
</target>

</project>
```

config.properties （Ant 运行监听器配置文件）

------Ant 配置资源文件，设置当出现错误时自动重新运行用例的次数------

retrycount=3   <!--设重运行 3 次，即用例失败时自动运行 3 次。如果 3 次后仍是 Failed，则该用例不通过，此功能可大大减少误报率。-->

sourcecodedir=src   <!--设置根目录 Src -->

sourcecodeencoding=UTF-8   <!--设置编码为中文-->

4.3　Android 和 iOS 自动化测试结果展示

自动测试用例测试结果记录样例（pass 或空），如图 4.4 所示。

测试用例编号	用例描述	测试步骤	测试对象名称描述	定位方式	控件元素	操作方法	测试数据	验证数据	延迟时间	测试结果
My_Login_001	正常登录	TS-001	我的一首页	By.name	我的	click				pass
My_Login_001	正常登录	TS-002	我的一登录页	By.name	我的专栏	click				pass
My_Login_001	正常登录	TS-003	我的一邮箱	By.name	邮箱	click				pass
My_Login_001	正常登录	TS-004	我的一用户名文本框	By.id	com.test.life:id/userName	sendkeys	7980068@qq.com			pass
My_Login_001	正常登录	TS-005	我的一密码文本框	By.id	com.test.life:id/password	sendkeys	13798359580			pass
My_Login_001	正常登录	TS-006	我的一登录按钮	By.name	登录	click		登录成功		pass
My_Logout_001	退出登录	TS-001	我的一首页	By.name	我的	click				
My_Logout_001	退出登录	TS-002	我的一下滑	By.driver	500	swipetoup				
My_Logout_001	退出登录	TS-003	我的一设置	By.id	com.test.life:id/setting	click				
My_Logout_001	退出登录	TS-004	我的一退出登录	By.name	退出登录	click		退出成功		

▲图 4.4

TestNG 自动测试报告样例如图 4.5 所示。

From: zouhui
Date: 2015-12-08 17:37
To: zouhui
Subject: autotest-ios3.3-smoke - Build # 39 - Successful!

测试总数：66

失败测试数量：0　　忽略测试数量：0

错误时日志：

　　All tests passed

运行场景：测试环境 | 冒烟测试用例 | 苹果iphone6手机 | IOS 9.0.1

测试报告详情：

Test	Methods Passed	Scenarios Passed	# skipped	# failed	Total Time	Included Groups	Excluded Groups
Test_Mail_Login	1	1	0	0	63.0 seconds		
Test_Post_AddDelModify	4	4	0	0	507.0 seconds		
Test_Post_Select	1	1	0	0	52.6 seconds		
Test_Community_Select	6	6	0	0	304.1 seconds		
Test_Community_AddDelModify	4	4	0	0	396.3 seconds		
Test_My_AddDelModify	7	7	0	0	417.7 seconds		

Class	Method	Authors	# of Scenarios	Running Counts	Parameters	Start	Time (ms)
Test_Mail_Login — passed							
logo.my.MyLoginAutoTest	My_Login_Mail_TC_001	unknown	1	1		1449563155213	41165
Test_Post_AddDelModify — passed							
logo.center.PostAddDelModifyAutoTest	CenterCharger_Picture_Post_TC_002	unknown	1	1		1449563369027	109659
	CenterCharger_SpecialText_Post_TC_004	unknown	1	1		1449563615288	87885
	CenterCharger_Text_Post_TC_001	unknown	1	1		1449563211268	141966
	CenterCharger_Video_Post_TC_003	unknown	1	1		1449563493084	107977
Test_Post_Select — passed							
logo.center.PostSelectAutoTest	CenterCharger_NotAuth_Post_TC_005	unknown	1	1		1449563717533	38137

▲图 4.5

第 5 章

API 接口自动化测试方案

5.1 概述

为什么做接口自动化测试？接口自动化测试的优势有哪些？

- 可以完成重复度较高的任务，减少重复工作量；
- 效率较高，需要更少的时间快速执行；可常用流程接口自动化，并定时构建自动执行；
- 及时发现后台 API 更新后的问题，保障主流程。

1．API 接口测试的概念

接口测试是测试系统组件间接口的一种测试。接口测试主要用于检测外部系统与系统之间以及内部各个子系统之间的交互点。测试的重点是检查数据的交换、传递和控制管理过程，以及系统间的相互逻辑依赖关系，包括处理的次数等。

我们首先要知道到底什么是被测对象接口，才能更好地对它进行测试。

2．什么是接口

API 接口是一种传输或操作数据的方式，广泛应用于 APP、服务端、Web 等，适用于数据的获取、更新、删除以及其他操作。较常见的就是 HTTP 接口和 WebService 接口，用得最多的是 HTTP 协议的 POST 接口和 GET 接口。

HTTP 请求包括 Headers、Url、Params、Body 值等，例如：

- 参数 1=值 2&参数 2=值 2
- https://passport.cnblogs.com/user/signin?ReturnUrl=http%3A%2F%2Fwww.cnblogs.com%2F

返回的数据基本都是 JSON 格式，例如：

- {"msg"：系统繁忙，请稍后再试"，"state"："3720001"}

3．接口测试的重点

（1）状态检查：请求是否正确，比如默认请求成功是 200 或 success，如果请求错误，则返回 404 等错误。

（2）检查返回数据的正确性与格式：Json 是一种常见的格式，也可是 xml 格式。

（3）边界和异常扩展检查：

- 参数字段默认值；
- 参数字段是否必填、是否为空检查；
- 参数字段携带错误值；
- 接口字段多少的检查：5 个字段变成 8 个字段等；
- 字段类型的检查：Int 型变成 String 型时如何判断；
- 限制条件：如店铺名重复，店铺标签修改重复，短信验证码次数超过 5 次，店铺名长度超过 20 等。

（4）流程接口测试：比如购物流程，依次要调用登录接口，商品加入购物车接口，提交订单接口，支付接口。同样要依照这些接口的逻辑流程进行接口测试，通常前一个接口会动态产生一个特定的数据关联到下一个接口。比如登录接口后会有特定的 token，供接下来的购物等接口调用，然后提交订单接口会产生一个特定的 orderid，供下一个支付接口调用。

常见的开源接口调试和抓包工具有 Postman 和 Fiddler，Mac 上使用 Charles。

常见的开源接口测试工具有 JMeter，具体参见第 9 章。

接口来源：一是开发人员提供 API 文档，二是 Fiddler 等工具抓包。

接口 Wiki 的文档样例如图 5.1 所示。

▲图 5.1

5.2 所用技术点

接口自动化测试所用技术点如下：

- Python 脚本；
- MySQL 数据库；
- Zentao（禅道）项目工具；
- Jenkins 持续集成；
- JDK、Eclipse、SVN 环境。

5.3 主要功能

- 接口自动化分层框架（用例集，测试数据，脚本）
- 接口测试用例覆盖买家 78 个、流程 12 个，未覆盖 2 个
- 接口用例覆盖卖家 61 个、流程 2 个，未覆盖 1 个
- 定时构建执行测试任务

- 自动邮件通知
- 支持版本更新迭代
- API 用例覆盖率的提高
- 自动化测试准确度的提高
- 代码维护的优化
- 版本更新的优化
- 测试结果报告展示无人值守情况下的自动测试

5.4 测试计划

时间计划：

对于有良好代码基础的熟手，可用一周时间做出演示 Demo。如果是从零开始的小白，则需建议用 3 到 6 个月的时间做出演示 demo。

对于有良好代码基础的熟手，可用一个月时间试运行测试用例。如果是从零开始的小白，则需建议用半年到一年试运行测试用例。

目前接口自动化框架计划

采用自动化框架的上下分层设计理念：上层管理整个自动化测试的开发、执行以及维护。在较为庞大的项目中，它体现出了重要的作用，可以管理整个自动测试，包括自动化测试用例执行的次序、测试脚本的维护，以及集中管理测试用例、测试报告和测试任务等。下层主要测试脚本的开发，充分使用相关测试工具，构建测试驱动，并完成测试业务逻辑等。

后续接口自动化框架计划

脚本更优化。目前流程类的接口脚本存在一些重复代码，需改进为根据录入 Zentao（禅道）的接口流程用例，即可以自动生成接口脚本，而不是重写一个有重复代码的脚本，提高脚本可维护性效率。目前脚本详见 7.4 节。

第 6 章

API 接口自动化环境搭建

6.1　Python 环境准备

编辑器：Eclipse + pydev 插件。

1．安装 Python

下载地址：http://www.Python.org/。

Python 有 Python 2 和 Python 3 两个版本，语法有些区别。

Python-2.7.10.amd64.msi 文件路径：https://pan.baidu.com/s/1skBPj0h

2．安装 Java JDK

下载地址：http://www.oracle.com/technetwork/Java/Javase/downloads/index.html。

3．安装 Eclipse

下载地址：http://www.eclipse.org/downloads/。

下载完成后，解压就可以直接使用，有的 Eclipse 版本不需要安装。

4．安装 pydev

（1）在线安装。

执行"Eclipse 中的 help→Install New Software"操作，如图 6.1 所示。

▲图 6.1

在 Work with 中输入 http://pydev.org/updates，如图 6.2 所示。

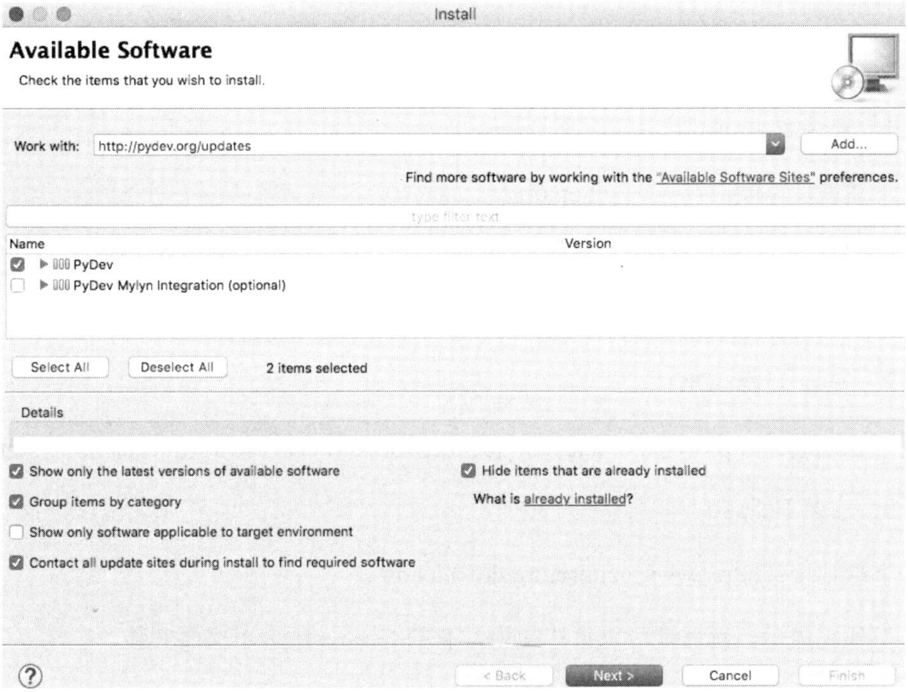

▲图 6.2

选择第一个框勾选，单击 Next 按钮，安装完成。

（2）离线安装。

下载 pydev 包，pydev 的版本号应与 Eclipse 版本号和 JDK 版本号相对应。将文件解压到 Eclipse 的 dropins 目录下，即文件夹 features 和 plugins，重启 Eclipse。安装文件路径：https://pan.baidu.com/s/1jH9vBmu。

（3）配置 pydev。

重启 Eclipse，执行"Eclipse 的 Window→Preferences→PyDev→Interpreters→Python interpreter"操作。单击"New"按钮后选择 Python 的安装地址：C:\Python27\python.exe，单击"Apply"按钮，安装成功。

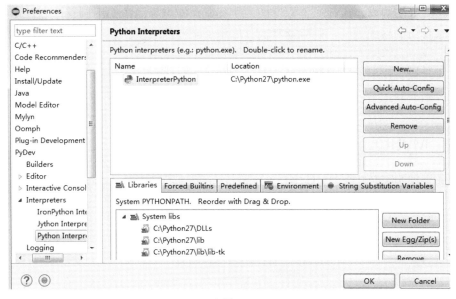

▲图 6.3

安装环境：安装接口资源 requests 包和数据库资源 MySQLdb 包。

（1）安装 Requests。

- 解压缩 Requests-2.11.1。
- 通过 Cmd 命令切换到相应目录文件下，执行 python setup.py install。

（2）安装 MySQLdb。

- 解压缩 MySQL-python-1.2.3c1。
- 通过命令 cmd 切换到相应目录文件下，执行 python setup.py install。

开始写 Python 代码，新建 Pydev 工程，写个 HelloWorld 程序并输出，环境安装后进行测试。

6.2　Zentao（禅道）项目管理工具

安装文件路径：https://pan.baidu.com/s/1eSDiVhK　　密码：bjbx

下载开源版本，进行一定程度的二次开发，即把界面中的用例管理等开发成符合测试接口的格式规范，同时读写数据库相关字段的数据。

工具提供接口测试用例管理、测试 bug 管理、版本关联测试用例的管理、多版本测试结果的记录，以及多个产品的接口测试管理，等等。

6.3　MySQL 数据库

客户端 Navicat 安装文件路径：https://pan.baidu.com/s/1slb8boh。

打开客户端建立连接，输入正确的本机 IP 地址、端口号、用户名 root、密码 test123456，如图 6.4 所示。

▲图 6.4

单击连接测试，如果正确，则会弹出提示：连接成功。双击即可看到表结构，如图 6.5 所示。

▲图 6.5

如果是官网下载的，则需要开启 root 用户的远程权限，并设置 root 密码，操作如下。

开放 Root 用户的远程权限和数值 Root 密码。

（1）Zentao（禅道）数据库端口号不能和已有端口号重复，否则会出现冲突并报错，默认 3306。

（2）在 Bin 目录下打开 MySQL 登录后，通过命令 cmd 切换到禅道安装目录 MySQL 的 bin 目录：

C:\Users\finer\Desktop\xamppdemo\mysql\bin>

输入 mysql -uroot -ptest123456，然后按回车键。

（3）设置 Root 远程权限及默认密码，比如为 root：

GRANT ALL PRIVILEGES ON *.* TO root@"%" IDENTIFIED BY "root";
flush privileges;

（4）修改 Root 密码为 test123456：

```
Mysql>Use mysql
Mysql>update use set password=PASSWORD('test123456') where user='root';
Mysql>flush privileges;
```

6.4 Fiddler 接口抓包工具

1. 安卓手机和 Fiddler 连接

Fiddler.exe 的下载地址：

http://www.telerik.com/download/fiddler。

做开发需要抓取手机 App 的 HTTP/HTTPS 数据包，当你想看看 App 发出的 HTTP 请求和响应是什么时，就需要抓包了。这样还可以得到一些不为人知的 API，比如在微信中，发红包才能看照片，就可以通过接口抓包成功破解，即不用发红包也可看照片。

所需工具：Fiddler 抓包软件。

下面介绍它的使用步骤。

（1）在 PC 上安装 Fiddler

下载地址：Fiddler.exe，http://www.telerik.com/download/fiddler。

（2）Fiddler 配置（配置完成后记得要重启 Fiddler）。

下载完成后安装，安装过程不再赘述。

1）选择下载正确的 fiddler 版本，单击启动—>帮助中—>About Fiddler，查看版本号，如图 6.6 所示。

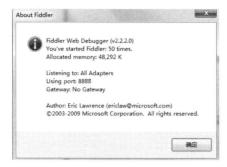

▲图 6.6

2）Fiddler 选项如图 6.7 所示。

▲图 6.7

3）在常规选项卡中，勾选允许远程计算机连接，如图 6.8 所示。

▲图 6.8

4）单击"连接"按钮，勾选"代理选项"、端口号"8888"，单击"OK"按钮，至此代理设置就完成了。注意，一定要重启软件，配置才会生效。

▲图 6.9

5）运行 cmd 命令，查看电脑的 IP 地址，如图 6.10 所示。

▲图 6.10

6）安卓手机一台，手机端代理设置，以华为手机为例。

- 如图 6.11 所示，找到你的 Wifi，必须让电脑和手机处于同一个 Wifi 下。
- 打开安卓手机，设置→WLAN→连接的 Wifi。
- 长按 Wifi 热点，选择修改网络，勾选显示高级选项。
- 代理设置为：手动；代理主机名为你的电脑 IP，端口就是刚才 Fiddler 设置的端口 8888，单击"保存"按钮，如图 6.11 所示。

第 6 章　API 接口自动化环境搭建 | 89

▲图 6.11

7）重启 Fiddler 以及手机 App 应用程序，即可在 Fiddler 界面看到手机请求数据和响应数据。左侧是主机 URL；右侧是嗅探到的 Headers，可查看头信息，WebForms 可查看参数名和值，TextView 可查看接口响应数据等，如图 6.12 所示。

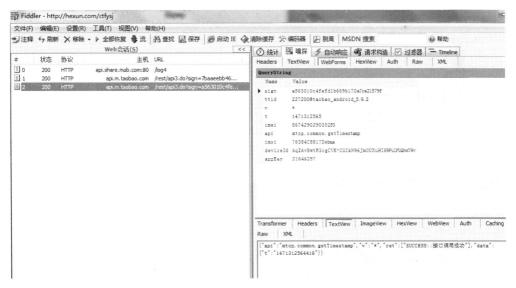

▲图 6.12

8）如果只看相应的 App 的数据，那么可以在 Fiddler 设置中勾选过滤器，使用过滤器，仅仅显示如图 6.13 所示的主机。

▲图 6.13

2．iOS 手机和 Fiddler 连接

Fiddler 不但能截获各种浏览器发出的 HTTP 请求，还能截获各种智能手机发出的 HTTP/HTTPS 请求。Fiddler 能捕获 iOS 设备发出的请求，比如 iPhone、iPad、MacBook 等苹果设备。同理，也可以截获 Andriod、Windows Phone 等设备发出的 HTTP/HTTPS 请求。最关键的是，对 iOS 应用抓包时可直接在 Windows PC 上进行，而并非一定要在 MAC 上。

3．iPhone 手机配置

（1）网络代理设置

安装 Fiddler 的机器与 iPhone 必须在同一个网络里，否则 iPhone 不能把 HTTP 请求发送到 Fiddler 的机器上来。一般情况下，我们手边上是台式机和手机，台式机只有网线，没有无线 Wifi，所以和手机不在同一个网络，这时，我们就需要设置代理了。

打开 iPhone，找到你的网络连接，打开 HTTP 代理，输入 Fiddler 所在机器的 IP 地址以及 Fiddler 的端口号 8888，如图 6.14 所示。

▲图 6.14

（2）Fiddler 证书安装

这一步是为了让 Fiddler 能捕获 HTTPS 请求。如果你只需截获 HTTP 请求，则可以忽略这一步。

1）首先要知道 Fiddler 所在机器的 IP 地址：这里安装了 Fiddler 的机器 IP 地址是：192.168.58.79。

2）打开 iPhone 的 Safari，访问 http://192.168.58.79:8888，单击"FiddlerRoot certificate"，然后安装证书，如图 6.15 所示。

▲图 6.15

大功告成,可以抓包了。iPhone 上的 App 发出的所有 HTTP/HTTPS 请求都可以被 Fiddler 获取。打开手机上的 App,看看 Fiddler 能否捕获。

(3) Fiddler 网络限速。

Fiddler 还为我们提供了一个很方便的网络限速功能,通过网络限速,我们可以模拟用户的一些真实环境。Fiddler 提供了网络限速的插件,具体不再赘述。官网下载地址:http://www.telerik.com/fiddler/add-ons。

(4) 注意事项。

用完了之后要退出 Fiddler,并把 iPhone 上的 Fiddler 代理关闭,以免 iPhone 或电脑连不上网。

如果只能捕获 HTTP 请求,而不能捕获 HTTPS 请求,其原因很可能是证书没有安装好。解决办法是,尝试重新安装 Fiddler 证书:首先删除 iPhone 上的 Fiddler 证书("设置→通用→描述文件"),然后再重新安装。

(5) Charles 抓包。

在 Mac 上使用功能与 Fiddler 类似的抓包工具，使用苹果 Mac 电脑，用 Charles 代替 Fidder，本书不对 Charles 进行介绍。

6.5　Postman 接口测试工具

第一步　安装：在 Chrome 浏览器中添加 Postman 插件、crx 文件和 rar 文件，然后解压缩，在 Chrome 扩展工具中加载解压缩后的文件，提示启动成功。注意，相应的谷歌版本和 Post 的 crx 版本要对应。

下载链接：http://pan.baidu.com/s/1cBmo2a。

（1）解压缩到 postman-4.1.2.rar 文件夹。

（2）在谷歌浏览器→工具→扩展工具→扩展程序中选中开发者模式，加载已解压的扩展程序选中 1 中的 postman-4.1.2.rar 文件夹，确定后，看到已启用勾选状态即正常。

（3）单击谷歌浏览器的应用，看到 PostMan，如图 6.16 所示。

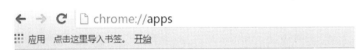

▲图 6.16

（4）单击 PostMan→REST Client，进入 Postman 接口测试页面，如图 6.17 所示。

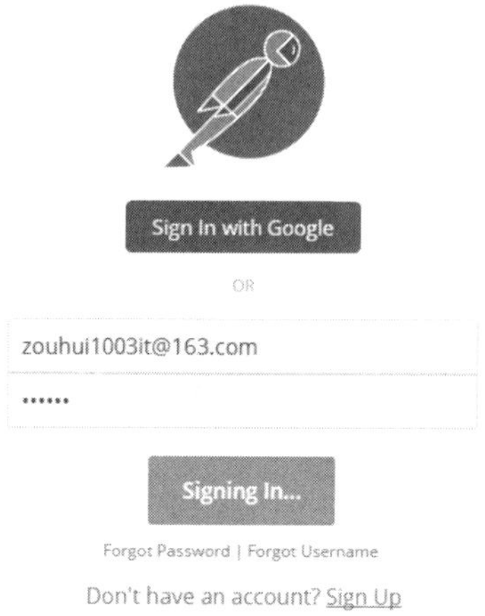

▲图 6.17

（5）．输入 get 接口请求地址，参数名，参数值，分别为

https://passport.cnblogs.com/user/signin，zouhui，123456

点击 Send，会看到服务端的返回的响应数据，点击保存到 Collections

▲图 6.18

（6）选择右边 SNIPPETS 下面第三行 Response body:Contains strin，在文本框中看到 Tests 下面的脚本，输入"百度一下"进行 body 校验；单击 Send 按钮，会看到运行结果显示为绿色 PASS body matches string，如图 6.19 所示。

▲图 6.19

说明：Postman 是接口测试中较为强大的工具，本书只做简单的介绍和使用，不做详细的讲解。

第 7 章

API 接口自动化源代码

7.1 基于 Python 的接口自动化脚本解析

1）python testAPIbuyerV1_2.py　　　　　买家接口集

2）python OrderSubmitFlowV1_2.py　　　提交订单流程接口

3）python OrderCancelFlowV1_2.py　　　取消订单流程接口

4）python testAPIbizV1_2.py　　　　　　卖家接口集

5）python bizOrderDetailV1_2.py　　　　卖家订单详情流程接口

6）python bizOrderDeliveryV1_2.py　　　卖家订单发货流程接口

7）python report.py　　　　　　　　　　买家和卖家接口测试报告

接下来，我们看看 API 测试自动化框架的脚本。

（1）界面接口测试用例

程序清单 7-1　Python 脚本

testAPIbuyerV1_2.py　　　　　　　　　买家接口集脚本如下

```
# -*- coding:utf-8 -*-                # 设置中文编码
import requests, MySQLdb, time, re    # 引入 request 包、MySQLdb 包、time 和 re 包接口请求相关的
                                      # 操作，数据库相关的操作，时间等相关的操作
import urllib, urllib2                # 引入 urllib 包，接口相关的操作


HOSTNAME = '192.168.215.55'           # 公共变量 主机 IP

def readSQLcase():                    # 读取测试用例，循环遍历相应项目的所有测试用例，读
                                      # 取产品 id=1，即买家的所有独立用例
    sql="SELECT t2.`case`,`title`,t2.`desc`,`precondition`,t2.`expect`,`keywords`,t2.`version` from
`zt_case`,`zt_casestep` as t2 inner join (select zt_casestep.`case`, max(version) as tv from `zt_casestep`  group by
zt_casestep.`case`) as t1 on t1.`case`=t2.`case` and t1.tv=t2.version ,`zt_testrun`,`zt_testtask` where zt_case.id =
t2.`case` and zt_case.type='interface' and zt_case.product=1 and zt_case.id=zt_testrun.`case` and
zt_testrun.task=zt_testtask.id and zt_testtask.id=8 and zt_case.status='investigate' and zt_case.id=157 "
    coon = MySQLdb.connect(user='root',passwd='test123456',db='zentao',port=3306,host='192.168.58.79',
```

```python
                charset='utf8')
                                        # 连接禅道 MySQL 数据库，host 一定要为 IP 地址
    cursor = coon.cursor()
    aa=cursor.execute(sql)              # 将执行上条 SQL 语句
    info = cursor.fetchmany(aa)
    for ii in info:
        case_list = []
        case_list.append(ii)
        GetToken()                      # 获取 token 的方法，为某用户登录后返回数据中的唯一标
                                        # 志，比如 13798359580 登录 token 值代表该用户 1 已登录，
                                        # 当 13798359581 登录时，token 值代表用户 2 已登录，这
                                        # 两个用户的 token 值必然是不同且唯一的 time.sleep(1)
        interfaceTest(case_list)        # 循环读取 SQL 中的一条测试用例
    coon.commit()
    cursor.close()
    coon.close()                        # 关闭数据库连接

def interfaceTest(case_list):           # 读取一条接口测试用例
    res_flags = []
    request_urls = []
    responses = []
    for case in case_list:
        try:
            case_id = case[0]           # 读取用例的关键信息用例 id
            interface_name = case[1]    # 读取用例的关键信息接口名称      对应标题
            method = case[3]            # 读取用例的关键信息接口方法      对应方法
            url = case[5]               # 读取用例的关键信息接口 url 地址  对应 keyword
            param = case[2]             # 读取用例的关键信息接口参数列表   对应参数
            res_check = case[4]         # 读取用例的关键信息接口返回校验   对应预期结果
        except Exception,e:
            return '测试用例格式不正确！%s'%e
        if param== '':
            new_url = 'http://'+HOSTNAME+url   # 如果接口没有参数，则接口设为 hostname+url
        elif param== 'null':
            new_url = 'http://'+HOSTNAME+url   # 如果接口参数为 null，则接口设为 hostname+url 。注意，
                                               # 当接口没有参数时，可在页面中输入 null 来判断是否有参数
        else:
            new_url = 'http://'+HOSTNAME+url+'?'+urlParam(param)   #如果接口有参数，则接口设为
```

```python
        request_urls.append(new_url)                                    hostname+url+param
if method.upper() == 'GET':                                     # 如果为 GET 方法，则读取如下 GET
                                                                # 请求和返回数据

        print str(case_id)+' ' +new_url                         # 打印用例 id 和接口地址
        headers = {'Host':HOSTNAME,
                    'Connection':'keep-alive',
                    'token':token,
                    'Content-Type': 'application/x-www-form-urlencoded',
                    'User-Agent': 'Apache-HttpClient/4.2.6 (java 1.5)'}  # 设置 http 头信息，包括主机名和
                                                                # 定 13798359580 用户登 token 值
        data = None
        results = requests.get(new_url,data,headers=headers).text  # 发送 get 请求，results 得到请求的
                                                                # 返回数据
        responses.append(results)
        res = readRes(results,res_check)                        # 对请求的返回数据进行校验，采用
                                                                # 正则表达式校验，校验结果有三种
                                                                # （pass、 fail、jFIF）
        if 'pass' == res:
            writeResult(case_id,'pass')                         # 写结果为 pass 到这个关联用例 id
            res_flags.append('pass')
            if JFIF(results):
                results = 'JFIF ok'                             # 校验 JFIF 则为图片
            else:
                print u'接口名称: '+interface_name               # 打印接口名称
                print u'接口地址: '+new_url                      # 打印接口地址
                print u'响应数据: '+results                      # 打印接口响应数据，打印接口 id
                                                                # 和返回的 success
                print str(case_id)+'-------------------------------------'+'success'+'----------------------------------------'
                continue
            print u'接口名称: '+interface_name                   # 打印接口名称
            print u'接口地址: '+new_url                          # 打印接口地址
            print u'响应数据: '+results                          # 打印接口响应数据，打印接口 id
                                                                # 和返回的 success
            print str(case_id)+'-------------------------------------'+'success'+'----------------------------------------'
        else:
            res_flags.append('fail')
            writeResult(case_id,'fail')                         #写结果为 fail 到这个关联用例 id
```

```
                if reserror(results):
                    writeBug(case_id,interface_name,new_url,"api response is error",res_check)
#如果是接口响应异常，即服务器异常时，这种方式即直接打印出错信息记录 bug 写到数据库
                else:
                    writeBug(case_id,interface_name,new_url,results,res_check)
#如果是接口校验数据错误，则用这种方式即把该接口的请求和响应数据记录 bug，并写到数据库
                print u'接口名称: '+interface_name            #打印接口名称
                print u'接口地址: '+new_url                   #打印接口地址
                print u'响应数据: '+results                   #打印接口响应数据， 打印接口 id
                                                             #和返回的 fail
                print str(case_id)+'----------------------------------------------'+'fail'+'----------------------------------------'

        else:                       #如果不是 GET 方法，则读取如下 POST 请求和返回数据，以下注释
                                    #部分与上面 GET 完全一致，仅把 GET 换成 POST 即可，其他都一样

            headers = {'Host':HOSTNAME,
                       'Connection':'keep-alive',
                       'token':token,
                       'Content-Type': 'application/x-www-form-urlencoded',
                       'User-Agent': 'Apache-HttpClient/4.2.6 (java 1.5)'
                       }
            data = None
            results = requests.post(new_url,data,headers=headers).text

responses.append(results)
            res = readRes(results,res_check)
            if 'pass' == res:
                writeResult(case_id,'pass')
                res_flags.append('pass')
                if JFIF(results):
                    results = 'JFIF ok'
                else:
                    print u'接口名称: '+interface_name
                    print u'接口地址: '+new_url
                    print u'响应数据: '+results
                    print str(case_id)+'------------------------------------'+'success'+'--------------------------------------'
                continue
            print u'接口名称: '+interface_name
            print u'接口地址: '+new_url
```

```
                print u'响应数据：'+results
                print str(case_id)+'----------------------------------------'+'success'+'----------------------------------------'
            else:
                res_flags.append('fail')
                writeResult(case_id,'fail')
                if reserror(results):
                    writeBug(case_id,interface_name,new_url,"api response is error",res_check)
                else:
                    writeBug(case_id,interface_name,new_url,results,res_check)
                print u'接口名称：'+interface_name
                print u'接口地址：'+new_url
                print u'响应数据：'+results
                print str(case_id)+'----------------------------------------'+'fail'+'----------------------------------------'

def readRes(res,res_check):                       # 校验结果。如一致则返回 pass，否则返回错误提示
    res = res.replace('":"','=').replace('":','=')    # 校验时替换符号为=号，再进行校验
    res_check = res_check.split(';')
    for s in res_check:
        if s in res:
            pass
        else:
            return  u'错误，返回参数和预期结果不一致'+str(s)
    return 'pass'

def urlParam(param):                              # 参数值的替换
    param1=param.replace('*','&')                 # 如果参数在数据库中为*,则替换为&
    param2=param1.replace('&quot;','\"')          # 如果参数在数据库中为&quot;，则替换成"，这是
                                                  # 因为在页面中输入的"，存储到数据库中就变成了
                                                  # &quot;，所以要转换
    return param2.replace(';','&')                # 如果参数在数据库中为;，则替换成&

def GetToken():                                   # 取用户登录的 token 值
    global token                                              # 定义 token 全局变量
    url = 'http://'+HOSTNAME+'/buyer/user/login.do'           # 接口的 URL
    params = {
        'phone': '13798359580',
        'pwd': '57ec2dd791e31e2ef2076caf66ed9b79',
    }                                                         # 参数为登录手机号和密码
```

```python
request = urllib2.Request(url = url, data = urllib.urlencode(params))    # 发送接口请求 URL 和参数
response = urllib2.urlopen(request)                                       # 返回响应数据
data = response.read()                                                    # 返回响应数据
regx = '.*"token":"(.*)","ud"'                                            # 正则表达式 token，左匹配"token":"，右匹配","ud"',
                                                                          # 即从 13798359580 登录后的响应数据中去进行
                                                                          # 匹配这个值，这个 toekn 值可在头信息中发送
                                                                          # 及取得，表示用户 13798359580 是登录状态

pm = re.search(regx, data)                                                # 取 token 匹配值
token = pm.group(1)                                                       # 如果匹配到则返回 token 值
regy = r'"state":(\d+)}'                                                  # 正则表达式 state，左匹配"state":，右匹配}
pn = re.search(regy, data)                                                # 如果匹配到则返回 state 值
state = pn.group(1)                                                       # 匹配第一个 state 的值
if state == '0':                                                          # 如果匹配到 state 值为 0，则返回 True；因为当登录
                                                                          # 成功会返回 state 匹配到 0，否则登录失败
    return True
return False

def reserror(results):                                                    # 接口页面返回匹配到的 HTML 页面服务器
                                                                          # 没有响应时，则为改调用接口错误

    global html
    regx = 'html'                                                         # regx 变量赋值为 HTML 字符串，如果服务器异常时会
                                                                          # 返回 404 等 HTML 标示，进行匹配

    pm = re.search(regx, results)
    if pm:
        return regx
    return False

def JFIF(results):                                                        # 接口页面返回匹配到为 JFIF 时，说明服务器
                                                                          # 返回为图片格式的乱码，表示该接口返回图
                                                                          # 片正确，否则错误

    global JFIF
    regx = 'JFIF'
    pm = re.search(regx, results)
    if pm:
        return regx
    return False
```

```python
def writeResult(case_id,result):                                    # 写测试结果到数据库
    result = result.encode('utf-8')
    now = time.strftime("%Y-%m-%d %H:%M:%S")                        # 当前时间格式话为此形式
    sql = "UPDATE zt_testrun set lastRunResult= %s, lastRunDate= %s, lastRunner='auto' where zt_testrun.task=8 and zt_testrun.`case`=%s;"
    param = (result,now,case_id)                                    # 把测试结果、时间、用例 id，作为动态
                                                                    # 参数写到数据库
    coon = MySQLdb.connect(user='root',passwd='test123456',db='zentao',port=3306,host='192.168.58.79',charset='utf8')
                                                                    # 连接禅道 MySQL 数据库，host 一定要为 IP 地址
    cursor = coon.cursor()
    cursor.execute(sql,param)                                       # 执行带参数的 SQL 语句
    coon.commit()
    cursor.close()
    coon.close()                                                    # 关闭数据库连接

def writeBug(bug_id,interface_name,request,response,res_check):     # 写测试 bug 到数据库
    interface_name = interface_name.encode('utf-8')                 # 接口名称字段格式转码为中文
    res_check = res_check.encode('utf-8')                           # 校验字段格式转码为中文
    response = response.encode('utf-8')                             # 接口响应数据字段格式转码为中文
    request = request.encode('utf-8')                               # 接口请求数据字段格式转码为中文
    now = time.strftime("%Y-%m-%d %H:%M:%S")                        # 写按此格式的当前时间
    bug_title = str(bug_id) + '_' + interface_name + '_出错了'       # 写 bug 的标题格式和内容信息
    step = '[请求报文]'+request+'<br/>'+'[预期结果]'+res_check+'<br/>'+'[响应报文]'+response
    # 写 bug 的重现步骤，即写出错接口的详 情到数据库
    sql = "INSERT INTO `zt_bug` (`openedDate`, `openedBy`, `lastEditedDate`, `lastEditedBy`, `status`,"\
    "`assignedTo`,`assignedDate`, `title`, `keywords`, `hardware`, `openedBuild`, `testtask`, `mailto`,"\
    "`steps`, `storyVersion`, `resolvedDate`, `resolvedBy`, `closedDate`, `closedBy`, `linkBug`, `case`,"\
    "`result`, `module`, `confirmed`, `resolution`, `duplicateBug`, `product`, `activatedCount`, `pri`,"\
    "`severity`) VALUES ('%s', '邹辉', '%s', '1', 'Active', "\
    "'邹辉', '%s', '%s', 'autotest', '1', '1', '1', '系统管理员', "\
    "'%s','1', '%s' , '邹辉', '%s', 'admin', '', '1', "\
    "'1', '1', '1', 'admin', '0', '1', '0', '1', "\
    "'1');"%(now,now,now,bug_title,step,now,now)
    #SQL 语句，其中，%s 为参数变量时间，bug_title=bug 标题，step=重现步骤，其他的字段名和值——对应
    coon = MySQLdb.connect(user='root',passwd='test123456',db='zentao',port=3306,host='192.168.58.79',charset='utf8')
    #连接禅道 MySQL 数据库，host 一定要为 IP 地址
```

```
    cursor = coon.cursor()
    cursor.execute(sql)                    # 执行上述 SQL 语句，并写 bug 到数据库
    coon.commit()
    cursor.close()
    coon.close()                           # 关闭数据库连接

if __name__ == '__main__':
    readSQLcase()                          # 执行 SQL 中的测试用例以及相关的操作
    print 'Done!'
```

（2）界面接口测试用例，提交订单流程接口，如图 7.1 所示。

▲图 7.1

程序清单 7-2　Python 脚本

python OrderSubmitFlowV1_2.py   提交订单流程接口脚本如下

```
# -*- coding:utf-8 -*-
import requests, xlrd, MySQLdb, time, sys, re
import urllib, urllib2, zlib
from _mysql import result
from httplib import ResponseNotReady

HOSTNAME = '192.168.215.55'            # 公共变量 主机 IP

def readSQLcase():                     # 提交订单流程的相关接口，循环遍历相应项目的所有接口，
                                       # 读取产品 id=1
    sql="SELECT t2.`case`,`title`,t2.`desc`,`precondition`,t2.`expect`,`keywords`,t2.`version` from
`zt_case`,`zt_casestep` as t2 inner join (select zt_casestep.`case`, max(version) as tv from `zt_casestep`　group by
```

```
zt_casestep.`case`) as t1 on t1.`case`=t2.`case` and t1.tv=t2.version where zt_case.id = t2.`case` and
zt_case.type='config' and zt_case.id=93 "
    coon = MySQLdb.connect(user='root',passwd='test123456',db='zentao',port=3306,host='192.168.58.79',
charset='utf8')
    #连接禅道 MySQL 数据库，host 一定要为 IP 地址
    cursor = coon.cursor()
    aa=cursor.execute(sql)              # 将执行上条 SQL 语句
    info = cursor.fetchmany(aa)
    for ii in info:
        case_list = []
        case_list.append(ii)
        GetToken()                      # 读取登录 token 值
        interfaceTest(case_list)        # 循环读取 SQL 中一条测试用例
    coon.commit()
    cursor.close()
    coon.close()                        # 关闭数据库连接

def interfaceTest(case_list):           # 读取一条接口测试用例的内容
    res_flags = []
    request_urls = []
    responses = []
    strinfo = re.compile('{preOrderSN}')  # 判断参数中如果有动态参数{preOrderSN}，则在代码中取值，作
                                         # 为订单号
    for case in case_list:
        try:
            case_id = case[0]            # 读取用例的关键信息用例 id
            interface_name = case[1]     # 读取用例的关键信息接口名称           对应标题
            method = case[3]             # 读取用例的关键信息接口方法           对应方法
            url = case[2]                # 读取用例的关键信息接口 url 地址       对应 kyword
            param = case[4]              # 读取用例的关键信息接口参数列表       对应参数
            res_check = case[5]          # 读取用例的关键信息接口返回校验       对应预期结果
        except Exception,e:
            return '测试用例格式不正确！%s'%e
        if param== '':
            new_url = 'http://'+HOSTNAME+url      # 如果接口没有参数，则接口设为 hostname+url
        elif param== 'null':
            new_url = 'http://'+HOSTNAME+url
# 如果接口参数为 null，则接口设为 hostname+url。注意，当接口没有参数时，在页面中输入 null 来判断是
# 否有参数
```

```python
            else:
                param = strinfo.sub(preOrderSN,param)
# 如有动态关联参数，则用值替换变量参数名，即把返回值传给此预提交订单号
            new_url = 'http://'+HOSTNAME+url+'?'+urlParam(param)
# 有参数接口则设为 hostname+url+param
            request_urls.append(new_url)
            if method.upper() == 'GET':               # 如果为 GET 方法，则读取如下 GET 请求和返回数
                print str(case_id)+' '+new_url        # 打印用例 id 和接口地址
                headers = {'Host':HOSTNAME,
                           'Connection':'keep-alive',
                           'token':token,
                           'Content-Type': 'application/x-www-form-urlencoded',
                           'User-Agent': 'Apache-HttpClient/4.2.6 (java 1.5)'}
# 设置 HTTP 头信息，包括主机名和用户登录的 token 值等
                data = None
                results = requests.get(new_url,data,headers=headers).text
# 发送 GET 请求，results 得到请求的返回数据
                responses.append(results)
                res = readRes(results,res_check)
# 对请求的返回数据进行校验，采用正则表达式校验，校验结果有三种：pass、fail、JIFI
                print results
                if 'pass' == res:
                    writeResult(case_id,'pass')              # 写结果为 pass 到这个关联用例 id
                    res_flags.append('pass')
                else:
                    res_flags.append('fail')
                    writeResult(case_id,'fail')              # 写结果为 fail 到这个关联用例 id
                    writeBug(case_id,interface_name,new_url,results,res_check)
#如果是接口校验数据错误，则这种方式记录 bug 写到数据库，以及该接口的请求和响应数据
            else:
                print str(case_id)+' '+new_url               # 打印用例 id 和接口地址
                headers = {'Host':HOSTNAME,
                           'Connection':'keep-alive',
                           'token':token,
                           'Content-Type': 'application/x-www-form-urlencoded',
                           'User-Agent': 'Apache-HttpClient/4.2.6 (java 1.5)'
                           }
# 设置 HTTP 头信息，包括主机名和用户登录的 token 值
                data = None
```

```
                results = requests.post(new_url,data,headers=headers).text
# 发送 POST 请求，results 得到请求的返回数据
                responses.append(results)
                res = readRes(results,res_check)
# 对请求的返回数据进行校验，采用正则表达式校验 ，校验结果有三种：  pass、 fail、JIFI

                print results
                if 'pass' == res:
                    writeResult(case_id,'pass')                         # 将结果 pass 写到这个关联用例 id
                    res_flags.append('pass')
                else:
                    res_flags.append('fail')
                    writeResult(case_id,'fail')                         # 将结果 fail 写到这个关联用例 id
                    if reserror(results):
                        writeBug(case_id,interface_name,new_url,"api response is error",res_check)
# 如果是接口响应异常，即服务器异常时，用这种方式将 bug 信息写到数据库，同时记录该接口的请求
# 信息和响应数据
                    else:
                        writeBug(case_id,interface_name,new_url,results,res_check)
# 如果是接口校验数据错误，则用这种方式将 bug 信息写到数据库，同时记录该接口的请求信息和响应数据
                try:
                    preOrderSN(results)
#根据上面的接口生成的预提交订单号，取值，供下面的接口作为参数进行发送请求
                except:
                    print 'ok'


def readRes(res,res_check):                     # 如果校验结果一致，则返回 pass，否则返回错误提示
    res = res.replace('":"','=').replace('":','=')  # 校验时":"替换符号为=号，再进行校验
    res_check = res_check.split(';')
    for s in res_check:
        if s in res:
            pass
        else:
            return  '错误，返回参数和预期结果不一致'+str(s)
    return 'pass'

def urlParam(param):                                        # 参数值的替换
    param1=param.replace('*','&')                           # 如果参数在数据库中为*，则替换为&
```

```python
        return param1.replace('&quot;','\"')
        # 如果参数在数据库中为&quot;，则替换成"。这是因为在页面中输入"时，存储在数据库中就变成了
        # &quot;，所以要转换

def GetToken():                                             # 取用户登录的 token 值
    global token                                            # 定义 token 全局变量
    url = 'http://'+HOSTNAME+'/buyer/user/login.do'         # 接口的 URL
    params = {
        'phone': '13798359580',
        'pwd': '57ec2dd791e31e2ef2076caf66ed9b79',
    }                                                       # 参数为登录手机号和密码
    request = urllib2.Request(url = url, data = urllib.urlencode(params))   # 发送接口请求 URL 和参数
    response = urllib2.urlopen(request)                     # 返回响应数据
    data = response.read()                                  # 返回响应数据
    regx = '.*"token":"(.*)","ud"'                          # 正则表达式 toekn，左匹配 "token":",右匹配","ud"
    pm = re.search(regx, data)                              # 取 token 匹配值
    token = pm.group(1)                                     # 如果匹配到则返回 token 值
    regy = r'"state":(\d+)}'                                # 正则表达式 state，左匹配"state":    右匹配}
    pn = re.search(regy,data)                               # 如果匹配到则返回 state 值
    state = pn.group(1)                                     # 匹配第一个 state 的值
    if state == '0':                                        # 如果匹配到 state 值为 0，则返回 True。因为当登录
                                                            # 成功时会返回 state 匹配到 0，否则登录失败
        return True
    return False


def preOrderSN(results):                                    # 预提交订单，参数取动态值
    global preOrderSN
    regx = '.*"preOrderSN":"(.*)","toHome"'                 # 预提交订单取值的正则表达式，左匹配
"preOrderSN":"   右匹配"toHome"
    pm = re.search(regx, results)
    if pm:
        preOrderSN = pm.group(1).encode('utf-8')            # 如果匹配到，则转换为中文并返回值
        return preOrderSN
    return False

def reserror(results):                                      # 接口页面返回匹配到为 HTML 页面服务器没有响应
```

```python
                                                    # 时，则改调用接口错误
    global html
    regx = 'html'
    pm = re.search(regx, results)                   # regx 变量赋值为 HTML 字符串，如果服务器异常则会
                                                    # 返回 404 等 HTML 标示，进行匹配
    if pm:
        return regx
    return False

def writeResult(case_id,result):                    # 写接口测试结果到数据库，关联的测试任务 id 为 8
    result = result.encode('utf-8')
    now = time.strftime("%Y-%m-%d %H:%M:%S")
    sql = "UPDATE zt_testrun set lastRunResult= %s, lastRunDate= %s, lastRunner='auto' where zt_testrun.task=8 and zt_testrun.`case`=%s;"
    param = (result,now,case_id)
    print result
    coon = MySQLdb.connect(user='root',passwd='test123456',db='zentao',port=3306,host='192.168.58.79',charset='utf8')
                                                    # 连接禅道 MySQL 数据库，host 一定要为 IP 地址
    cursor = coon.cursor()
    cursor.execute(sql,param)
    coon.commit()
    cursor.close()
    coon.close()

def writeBug(bug_id,interface_name,request,response,res_check):    # 写测试 bug 到数据库
    interface_name = interface_name.encode('utf-8')                # 接口名称字段格式转码为中文
    res_check = res_check.encode('utf-8')
    response = response.encode('utf-8')
    request = request.encode('utf-8')
    now = time.strftime("%Y-%m-%d %H:%M:%S")
    bug_title = str(bug_id) + '_' + interface_name + '_出错了'       # 写 bug 的标题格式和内容信息
    step = '[请求报文]<br />'+request+'<br/>'+'[预期结果]<br/>'+res_check+'<br/>'+'<br/>'+'[响应报文]<br />'+'<br/>'+response
                                                    # 写 bug 的重现步骤等，即出错接口的详情到数据库
    sql = "INSERT INTO `zt_bug` (`openedDate`, `openedBy`, `lastEditedDate`, `lastEditedBy`, `status`,"\
    "`assignedTo`,`assignedDate`, `title`, `keywords`, `hardware`, `openedBuild`, `testtask`, `mailto`,"\
    "`steps`, `storyVersion`, `resolvedDate`, `resolvedBy`, `closedDate`, `closedBy`, `linkBug`, `case`,"\
    "`result`, `module`, `confirmed`, `resolution`, `duplicateBug`, `product`, `activatedCount`, `pri`,"\
    "`severity`) VALUES ('%s', '邹辉', '%s', '1', 'Active', "\
    "'邹辉', '%s', '%s', 'autotest', '1', '1', '1', '系统管理员', "\
```

```
        "'%s','1', '%s' , '邹辉', '%s', 'admin', '', '1', "\
        "'1', '1', '1', 'admin', '0', '1', '0', '1', "\
        "'1');"%(now,now,now,bug_title,step,now,now)
# 写 bug 的 SQL 语句，关联产品 id 为 1 时，即买家产品。其中，%s 为参数变量时间，bug_title=bug 标题，
# step=重现步骤，其他的字段名和值一一对应
    coon = MySQLdb.connect(user='root',passwd='test123456',db='zentao',port=3306,host='192.168.58.79',charset='utf8')
    #连接禅道 MySQL 数据库，host 一定要为 IP 地址
    cursor = coon.cursor()
    cursor.execute(sql)                    # 执行上述 SQL 语句，写 bug 信息到数据库
    coon.commit()
    cursor.close()
    coon.close()

if __name__ == '__main__':
    readSQLcase()                          # 执行 SQL 中的测试用例以及相关操作
    print 'Done!'
    time.sleep(60)
```

（3）界面接口测试用例，取消订单流程接口如图 7.2 所示。

▲图 7.2

程序清单 7-3 Python 脚本

python OrderCancelFlowV1_2.py   取消订单流程接口脚本如下
```
# -*- coding:utf-8 -*-
import requests, xlrd, MySQLdb, time, sys, re
import urllib, urllib2, zlib
```

```python
from _mysql import result
from httplib import ResponseNotReady

HOSTNAME = '192.168.215.55'                        # 公共变量 主机 IP

def readSQLcase():                                 # 取消订单流程的相关接口，循环遍历相应项目的所有接口
                                                   # 读取产品 id=1
    sql="SELECT t2.`case`,`title`,t2.`desc`,`precondition`,t2.`expect`,`keywords`,t2.`version` from `zt_case`,`zt_casestep` as t2 inner join (select zt_casestep.`case`, max(version) as tv from `zt_casestep`  group by zt_casestep.`case`) as t1 on t1.`case`=t2.`case` and t1.tv=t2.version where zt_case.id = t2.`case` and zt_case.type='config' and zt_case.id=198 "
    coon = MySQLdb.connect(user='root',passwd='test123456',db='zentao',port=3306,host='192.168.58.79',charset='utf8')
                                                   # 连接禅道 MySQL 数据库，host 一定要为 IP 地址
    cursor = coon.cursor()
    aa=cursor.execute(sql)                         # 将执行上条 SQL 语句
    info = cursor.fetchmany(aa)
    for ii in info:
        case_list = []
        case_list.append(ii)
        GetToken()                                 # 读取登录 token 值
        interfaceTest(case_list)                   # 循环读取 SQL 中一条测试用例
    coon.commit()
    cursor.close()
    coon.close()                                   # 关闭数据库连接

def interfaceTest(case_list):                      # 读取一条接口测试用例的内容
    res_flags = []
    request_urls = []
    responses = []
    strinfo = re.compile('{preOrderSN}')           # 判断参数中如果有动态参数{preOrderSN}，则在代码中取
                                                   # 值，对应作为这个预提交订单号
    orderinfo = re.compile('{ordersn}')            # 判断参数中如果有动态参数{ordersn}则在代码中取值，对
                                                   # 应作为这个订单号
    for case in case_list:
        try:
            case_id = case[0]                      # 读取用例的关键信息用例 id
            interface_name = case[1]               # 读取用例的关键信息接口名称      对应标题
            method = case[3]                       # 读取用例的关键信息接口方法      对应方法
```

```
        url = case[2]              # 读取用例的关键信息接口 URL 地址       对应 kyword
        param = case[4]            # 读取用例的关键信息接口参数列表       对应参数
        res_check = case[5]        # 读取用例的关键信息接口返回校验       对应预期结果
except Exception,e:
        return '测试用例格式不正确！%s'%e
if param== '':
        new_url = 'http://'+HOSTNAME+url      # 如果接口没有参数，接口设为 hostname+url
                                              # elif param== 'null':
        new_url = 'http://'+HOSTNAME+url      # 如果接口参数 null 接口设为
                                              # hostname+url 。注意，当接口没有参数 时，可在页
                                              # 面中输入 null 来判断有没有参数
else:
        param = strinfo.sub(preOrderSN,param) # 如有动态关联参数，则替换变量参数传值给此
                                              # preOrderSN 参数
        param = orderinfo.sub(ordersn,param)  # 如有动态关联参数，则替换变量参数传值给此
                                              # ordersn 参数
        new_url = 'http://'+HOSTNAME+url+'?'+urlParam(param)
                                              # 如果接口有参数，则接口设为 hostname+url+param
        request_urls.append(new_url)
if method.upper() == 'GET':                   # 如果为 GET 方法，读取如下 GET 请求和返回数
        print str(case_id)+' '+new_url        # 打印用例 id 和接口地址
        headers = {'Host':HOSTNAME,
                   'Connection':'keep-alive',
                   'token':token,
                   'Content-Type': 'application/x-www-form-urlencoded',
                   'User-Agent': 'Apache-HttpClient/4.2.6 (Java 1.5)'}
                                              # 设置 HTTP 头信息，包括主机名和用户登录的 token 值
        data = None
        results = requests.get(new_url,data,headers=headers).text
                                              # 发送 GET 请求，results 得到请求的返回数据
        responses.append(results)
        res = readRes(results,res_check)      # 对请求的返回数据进行校验、正则表达式校验。校验
                                              # 结果有三种：pass、fail、JIGF
        print results
        if 'pass' == res:
                writeResult(case_id,'pass')   # 写结果为 pass 到这个关联用例 id
                res_flags.append('pass')
        else:
```

```python
                res_flags.append('fail')
                writeResult(case_id,'fail')          # 写结果为 fail 到这个关联用例 id
                writeBug(case_id,interface_name,new_url,results,res_check)
                                                     # 如果是接口校验数据错误,则这种方式记录 bug 写
                                                     # 到数据库,#以及该接口的请求和响应数据
    else:
        print str(case_id)+' '+new_url
        headers = {'Host':HOSTNAME,
                   'Connection':'keep-alive',
                   'token':token,
                   'Content-Type': 'application/x-www-form-urlencoded',
                   'User-Agent': 'Apache-HttpClient/4.2.6 (Java 1.5)'
                   }                      # 设置 HTTP 头信息,包括主机名和用户登录的 token 值
        data = None
        results = requests.post(new_url,data,headers=headers).text  # 发送 post 请求,results 得到请#求
                                                                    # 的返回数据
        responses.append(results)
        res = readRes(results,res_check)             # 对请求的返回数据进行校#验、正
                                                     # 则表达式校验。校验结果有三种:
                                                     # pass、fail、JFIF
        print results
        if 'pass' == res:
            writeResult(case_id,'pass')              # 写结果为 pass 到这个关联用例 id
            res_flags.append('pass')
        else:
            res_flags.append('fail')
            writeResult(case_id,'fail')              # 写结果为 fail 到这个关联用例 id
            if reserror(results):
                writeBug(case_id,interface_name,new_url,"api response is error",res_check)
# 如果是接口响应异常,即服务器异常时,这种方式记录 bug 写到数据库,以及该接口的
# 请求和响应数据
            else:
                writeBug(case_id,interface_name,new_url,results,res_check)
#如果是接口校验数据错误,则这种方式记录 bug 写到数据库,以及该接口的请求和响应数据
        try:
            preOrderSN(results)         # 参数化动态数据、订单号,即根据上面的接口生成的
                                        # 预提交订单号,取值供下面接口作为参数进行发送请求
        except:
            print 'ok'
```

```python
            try:
                ordersn(results)              # 参数化动态数据订单号，即根据上面的接口生成的订单
                                              # 号，取值供下面接口作为参数进行发送请求
            except:
                print ''


def readRes(res,res_check):                   # 校验结果如一致，则返回 pass，否则返回错误提示
    res = res.replace('":"','=").replace('":','="')  # 校验时替换符号为=号，再进行校验
    res_check = res_check.split(';')
    for s in res_check:
        if s in res:
            pass
        else:
            return '错误，返回参数和预期结果不一致'+str(s)
    return 'pass'

def urlParam(param):                          # 参数值的替换
    param1=param.replace('*','&')             # 如果参数在数据库中为*，则代码
                                              # 替换为&
    return param1.replace('&quot;','\"')      # 如果参数在数据库中为;替换成&

def GetToken():                               # 取用户登录的 token 值
    global token                              # 定义 token 全局变量
    url = 'http://'+HOSTNAME+'/buyer/user/login.do'   # 接口的 URL
    params = {
        'phone': '13798359580',
        'pwd': '57ec2dd791e31e2ef2076caf66ed9b79',
    }                                         # 参数为登录手机号和密码
    request = urllib2.Request(url = url, data = urllib.urlencode(params))  # 发送接口请求 URL 和参数
    response = urllib2.urlopen(request)       # 返回响应数据
    data = response.read()                    # 返回响应数据
    regx = '.*"token":"(.*)","ud"'            # 正则表达式 toekn，左匹配 "token":"右匹配","ud"
    pm = re.search(regx, data)                # 取 token 匹配值
    token = pm.group(1)                       # 如果匹配到，则返回 token 值
    regy = r'"state":(\d+)}'                  # 正则表达式 state，左匹配"state":右匹配}
    pn = re.search(regy,data)                 # 如果匹配到则返回 state 值
```

```python
        state = pn.group(1)                          # 匹配第一个 state 的值
        if state == '0':                             # 如果匹配到 state 值为 0 则返回 True；因为当登录成
                                                     # 功会返回 state 匹配到 0，否则登录失败
            return True
    return False


def preOrderSN(results):                             # 预提交订单号参数取动态值
    global preOrderSN
    regx = '.*"preOrderSN":"(.*)","toHome"'          # 预提交订单号取值的正则表达式，左匹配
                                                     # "preOrderSN":"   右匹配"toHome"
    pm = re.search(regx, results)
    if pm:
        preOrderSN = pm.group(1).encode('utf-8')     # 如果匹配到，则转换为 UTF-8 格式并返回值
        return preOrderSN
    return False


def ordersn(results):                                # 订单号参数取动态值
    global ordersn
    regx = '.*"tradeNo":"(.*)"}'                     # 订单号取值的正则表达式，左匹配"tradeNo":"
                                                     # 右匹配}
    pm = re.search(regx, results)
    if pm:
        ordersn = pm.group(1).encode('utf-8')        # 如果匹配到则转换为 UTF-8 格式并返回值
        return ordersn
    return False


def reserror(results):                               # 接口页面返回匹配到为 HTML 页面服务器没有
                                                     # 响应时，则改调用接口错误
    global html
    regx = 'html'
    pm = re.search(regx, results)                    # regx 变量赋值为 HTML 字符串，如果服务器异常时会
                                                     # 返回 404 等 HTML 标示，进行匹配
    if pm:
        return regx
    return False


def writeResult(case_id, result):                    # 写接口测试结果到数据库，关联的测试任务 id 为 8
    result = result.encode('utf-8')
```

```python
        now = time.strftime("%Y-%m-%d %H:%M:%S")
        sql = "UPDATE zt_testrun set lastRunResult= %s, lastRunDate= %s, lastRunner='auto' where zt_testrun.task=8 and zt_testrun.`case`=%s;"
        param = (result,now,case_id)
        print result
        coon = MySQLdb.connect(user='root',passwd='test123456',db='zentao',port=3306,host='192.168.58.79',charset='utf8')
        #连接禅道 MySQL 数据库，host 一定要为 IP 地址
        cursor = coon.cursor()
        cursor.execute(sql,param)
        coon.commit()
        cursor.close()
        coon.close()

def writeBug(bug_id,interface_name,request,response,res_check):        # 写测试 bug 到数据库
        interface_name = interface_name.encode('utf-8')               # 接口名称字段格式转码为中文
        res_check = res_check.encode('utf-8')
        response = response.encode('utf-8')
        request = request.encode('utf-8')
        now = time.strftime("%Y-%m-%d %H:%M:%S")
        bug_title = str(bug_id) + '_' + interface_name + '_出错了'     # 写 bug 的标题格式和内容信息
        step = '[请求报文]<br />'+request+'<br/>'+'[预期结果]<br/>'+res_check+'<br/>'+'<br/>'+'[响应报文]<br />'+'<br/>'+response    #写 bug 的重现步聚等,即出错接口的详情到数据库
        sql = "INSERT INTO `zt_bug` (`openedDate`, `openedBy`, `lastEditedDate`, `lastEditedBy`, `status`,"\
              "`assignedTo`,`assignedDate`, `title`, `keywords`, `hardware`, `openedBuild`, `testtask`, `mailto`,"\
              "`steps`, `storyVersion`, `resolvedDate`, `resolvedBy`, `closedDate`, `closedBy`, `linkBug`, `case`,"\
              "`result`, `module`, `confirmed`, `resolution`, `duplicateBug`, `product`, `activatedCount`, `pri`,"\
              "`severity`) VALUES ('%s', '邹辉', '%s', '1', 'Active', "\
              "'邹辉', '%s', '%s', 'autotest', '1', '1', '1', '系统管理员', "\
              "'%s','1', '%s' , '邹辉', '%s', 'admin', '', '1', "\
              "'1', '1', '1', 'admin', '0', '1', '0', '1', "\
              "'1');"%(now,now,now,bug_title,step,now,now)
# 写 bug 的 SQL 语句，关联产品 id 为 1，即买家产品。其中，%s 为参数变量时间，bug_title=bug 标题，
# step=重现步骤，其他字段名和值一一对应
        coon = MySQLdb.connect(user='root',passwd='test123456',db='zentao',port=3306,host='192.168.58.79',charset='utf8')
#连接禅道 MySQL 数据库，host 一定要为 IP 地址
        cursor = coon.cursor()
        cursor.execute(sql)                                           # 执行上述 SQL 语句，写 bug 到数据
```

```
            coon.commit()
            cursor.close()
        coon.close()

if __name__ == '__main__':
    readSQLcase()                                  # 执行 SQL 中的测试用例以及相关的操作
    print 'Done!'
    time.sleep(60)
```

（4）界面接口测试用例。

程序清单 7-4　Python 脚本

```
python testAPIbizV1_2.py                           卖家接口集脚本如下
# -*- coding:utf-8 -*-
import requests, MySQLdb, time, re
import urllib, urllib2
from random import choice


HOSTNAME = '192.168.215.55'                        # 公共变量 主机 IP


def readSQLcase():                                 # 读取测试用例，循环遍历相应项目的所有测试
                                                   # 用例，读取产品 id=3，即卖家的所有独立用例
    sql="SELECT t2.`case`,`title`,t2.`desc`,`precondition`,t2.`expect`,`keywords`,t2.`version` from
`zt_case`,`zt_casestep` as t2 inner join (select zt_casestep.`case`, max(version) as tv from `zt_casestep`    group by
zt_casestep.`case`) as t1 on t1.`case`=t2.`case` and t1.tv=t2.version ,`zt_testrun`,`zt_testtask` where zt_case.id =
t2.`case` and zt_case.type='interface' and zt_case.product=3 and zt_case.id=zt_testrun.`case` and
zt_testrun.task=zt_testtask.id and zt_testtask.id=9 and zt_case.status='normal' "
    coon = MySQLdb.connect(user='root',passwd='test123456',db='zentao',port=3306,host='192.168.58.79',
charset='utf8')
    #连接禅道 MySQL 数据库，host 一定要为 IP 地址
    cursor = coon.cursor()
    aa=cursor.execute(sql)                         # 将执行上条 SQL 语句
    info = cursor.fetchmany(aa)
    for ii in info:
        case_list = []
        case_list.append(ii)
        GetToken()                                 # 获取 token，默认为该用户 token 信息进行登录
        interfaceTest(case_list)                   # 循环读取 SQL 中一条测试用例
    coon.commit()
    cursor.close()
```

```python
        coon.close()                              # 关闭数据库连接

def interfaceTest(case_list):                     # 读取一条接口测试用例
    res_flags = []
    request_urls = []
    responses = []
    storenameinfo = re.compile('{storename}')     # 判断参数中如果有动态参数{storename}，则在代码中取
                                                  # 值，对应作为这个店铺名称
    storetagnameinfo = re.compile('{storetagname}')  # 判断参数中如果有动态参数{storetagname}，则在代
                                                     # 码中取值，对应作为这个标签名称
    deltagidinfo = re.compile('{deltagid}')       # 判断参数中如果有动态参数{deltagid},则在代码数据
                                                  # 库中取值，对应作为这个tagid
    deltagnameinfo = re.compile('{deltagname}')   # 判断参数中如果有动态参数{deltagname}，则在代码
                                                  # 数据库中取值，对应作为这个标签名
    for case in case_list:
        try:
            case_id = case[0]                     # 读取用例的关键信息用例id
            interface_name = case[1]              # 读取用例的关键信息接口名称      对应标题
            method = case[3]                      # 读取用例的关键信息接口方法      对应方法
            url = case[5]                         # 读取用例的关键信息接口 URL 地址   对应 keyword
            param = case[2]                       # 读取用例的关键信息接口参数列表   对应参数
            res_check = case[4]                   # 读取用例的关键信息接口返回校验   对应预期结果
        except Exception,e:
            return '测试用例格式不正确！ %s'%e    # 抛异常处理
        if param=='':
            new_url = 'http://'+HOSTNAME+url      # 如果接口没有参数 接口设为 hostname+url
        elif param== 'null':
            new_url = 'http://'+HOSTNAME+url      # 如果接口参数为 null，则接口设为 hostname+url。注
                                                  # 意，当接口没有参数时，可在页面中输入 null 来判
                                                  # 断有没有参数
        else:
            param = storenameinfo.sub(storename,param)
#如有动态关联参数{storename}，则用值替换变量参数名，即把返回值传给此店铺名称
            param = storetagnameinfo.sub(storetagname,param)
#如有动态关联参数{storetagname}，则用值替换变量参数名，即把返回值传给此店铺标签名称
            param = deltagidinfo.sub(deltagid,param)
#如有动态关联参数{deltagid}，则用值替换变量参数名，即把返回值传给此店铺标签 id
            param = deltagnameinfo.sub(deltagname,param)
```

```
                #如有动态关联参数{deltagname}，则用值替换变量参数名，即把返回值传给此店铺标签名称
                    new_url = 'http://'+HOSTNAME+url+'?'+urlParam(param)
                #如果接口有参数，则接口设为 hostname+url+param
                    request_urls.append(new_url)
                if method.upper() == 'GET':                # 如果为 GET 方法，读取如下 GET 请求和返回数据
                    print str(case_id)+' '+new_url         # 打印用例 id 和接口地址
                    headers = {'Host':HOSTNAME,
                              'Connection':'keep-alive',
                              'token':token,
                              'Content-Type': 'application/x-www-form-urlencoded',
                              'User-Agent': 'Apache-HttpClient/4.2.6 (Java 1.5)'}    # 设置 HTTP 头信息，包括主机
                                                                                    # 名和用户登录的 token 值等

                    data = None
                    results = requests.get(new_url,data,headers=headers).text        # 发送 GET 请求，results 得到
                                                                                     # 请求的返回数据

                    responses.append(results)
                    res = readRes(results,res_check)                 # 对请求的返回数据进行校验、正则表达
                                                                     # 式校验。校验结果有三种：pass、fail、JFIF
                    print results.encode('utf-8')
                    if 'pass' == res:
                        writeResult(case_id,'pass')                  # 写结果为 pass 到这个关联用例 id
                        res_flags.append('pass')
                    else:
                        res_flags.append('fail')
                        writeResult(case_id,'fail')                  # 写结果为 fail 到这个关联用例 id
                        writeBug(case_id,interface_name,new_url,results,res_check)
# 如果是接口校验数据错误，则用这种方式记录 bug 并写到数据库，以及该接口的请求和    响应数据
            else:
# 如果不为 GET 方法，则读取如下 POST 请求和返回数据，以下注释部分与上面 GET 完全一致，把 GET
# 换成 POST 即可，其他都一样
                    headers = {'Host':HOSTNAME,
                              'Connection':'keep-alive',
                              'token':token,
                              'Content-Type': 'application/x-www-form-urlencoded',
                              'User-Agent': 'Apache-HttpClient/4.2.6 (Java 1.5)'
                              }           #设置 HTTP 头信息，包括主机名和用户登录的 token 值
                    data = None
                    results = requests.post(new_url,data,headers=headers).text
```

```
# 发送 POST 请求，results 得到请求的返回数据
            responses.append(results)
            res = readRes(results,res_check)
# 对请求的返回数据进行校验、正则表达式校验。校验结果有三种：pass、 fail、JFIF
            if 'pass' == res:
                writeResult(case_id,'pass')                    # 写结果为 pass 到这个关联用例 id
                res_flags.append('pass')
                if JFIF(results):
                    results = 'JFIF ok'                         # 校验 JFIF 则为图片
                else:
                    print u'接口名称: '+interface_name            # 打印接口名称
                    print u'接口地址: '+new_url                   # 打印接口地址
                    print u'响应数据: '+results                   # 打印接口响应数据 ,打印接口 id 和
                                                                # 返回 success
                    print str(case_id)+'----------------------------------------'+'success'+'----------------------------------------'
                    continue
                print u'接口名称: '+interface_name                # 打印接口名称
                print u'接口地址: '+new_url                       # 打印接口地址
                print u'响应数据: '+results                       # 打印接口响应数据 ,打印接口 id
                                                                # 和返回 success
                print str(case_id)+'----------------------------------------'+'success'+'----------------------------------------'
            else:
                res_flags.append('fail')
                writeResult(case_id,'fail')                    # 写结果为 fail 到这个关联用例 id
                if reserror(results):
                    writeBug(case_id,interface_name,new_url,"api response is error",res_check)
#如果是接口响应异常，即服务器异常时，则用这种方式即直接打印错误信息记录 bug 并写到数据库
                else:
                    writeBug(case_id,interface_name,new_url,results,res_check)
#如果是接口校验数据错误，则用这种方式及把改接口的请求和响应数据详细记录 bug 并写到数据库
                print u'接口名称: '+interface_name          # 打印接口名称
                print u'接口地址: '+new_url                 # 打印接口地址
                print u'响应数据: '+results                 # 打印接口响应数据，打印接口 id 和返回 fail
                print str(case_id)+'----------------------------------------'+'fail'+'----------------------------------------'

def readRes(res,res_check):                              # 校验结果如一致，则返回 pass，否则返回错误提示
    res = res.replace('"':"',"=").replace('":',"=")      # 校验时替换符号为=号，再进行校验
    res_check = res_check.split(';')
```

```python
    for s in res_check:
        if s in res:
            pass
        else:
            return '错误，返回参数和预期结果不一致'+str(s)
    return 'pass'

def urlParam(param):                                        # 参数值的替换
    param1=param.replace('*','&')                           # 如果参数在数据库中为* 替换为&
    param2=param1.replace('&quot;','\"')                    # 如果参数在数据库中为&quot;则
    #替换成"。这是因为在页面中输入"时，存储在数据库中就变成了&quot;，所以要转换
    return param2.replace(';','&')                          # 如果参数在数据库中为;则替换成&

def GetToken():                                             # 取用户登录的 token 值
    global token                                            # 定义 token 全局变量
    url = 'http://'+HOSTNAME+'/portal/user/login.do'        # 接口的 URL
    params = {
        'phone': '13798359581',
        'pwd': '57ec2dd791e31e2ef2076caf66ed9b79',
    }                                                       # 参数为登录手机号和密码
    request = urllib2.Request(url = url, data = urllib.urlencode(params))  # 发送接口请求 URL 和参数
    response = urllib2.urlopen(request)                     # 返回响应数据
    data = response.read()                                  # 返回响应数据
    regx = '.*"token":"(.*)","ud"'                          # 正则表达式 toekn，左匹配
                                                            # "token":"   ，右匹配","ud"'
    pm = re.search(regx, data)                              # 取 token 匹配值
    token = pm.group(1)                                     # 如果匹配到则返回 token 值
    regy = r'"state":(\d+)}'                                # 正则表达式 state，左匹配"state":
                                                            # 右匹配}
    pn = re.search(regy,data)                               # 如果匹配到则返回 state 值
    state = pn.group(1)                                     # 匹配第一个 state 的值
    if state == '0':
    # 如果匹配到 state 值为 0 ,则返回 True；因为当登录成功会返回 state 匹配到 0，否则登录失败
        return True
    return False

def reserror(results):              # 接口页面返回匹配到的 HTML 页面服务器，没有响应时，则改调用接口错误
```

```python
    global html
    regx = 'html'                        # 匹配 HTML 字符串
    pm = re.search(regx, results)
                                         # regx 变量赋值为 HTML 字符串，如果服务器异常，会返回 404 等 HTML
                                         # 标示，进行匹配
    if pm:
        return regx
    return False

def JFIF(results):
#接口页面返回匹配到为 JFIF 时，说明服务器返回为图片格式的乱码，表示该接口返回图片正确，否则错误
    global JFIF
    regx = 'JFIF'                        # 匹配 JFIF 字符串
    pm = re.search(regx, results)        # 跟返回的响应数据进行匹配，是否存在 JFIF
    if pm:
        return regx                      # 如果匹配到，则返回改 JFIF 字符串
    return False

def storename(param):                    # 动态店铺名，tstore+动态时间戳作为店铺名参数，以免重名
    global storename                     # 定义店铺名称全局变量
    storename='tstore'+str(int(time.time()))   # 设置店铺名称为 tstore+数字型时间戳
    return storename                     # 返回店铺名称

def storetagname(param):                 # 店铺标签名为有 8 位随机字母，作为参数，以免重复
    global storetagname                  # 定义店铺标签名称全局变量
    storetagname = ''.join([choice('AaBbCcDdEeFfGgHhIiJjKkLlMmNnOoPpQqRrSsTtUuVvWwXxYyZz') for i in range(8)])
                                         # 店铺标签名称在字母中随机取 8 个字符
    return storetagname                  # 返回店铺标签名称

def deltagid(param):                     # 获取要删除的 tagid  根据数据库中的 id 值作为参数
    sql = "select id from t_tag_info where shopid=1000169 and status=1"
                                         # 根据参数，在所测系统的数据库中查找相应的 id
    coon = MySQLdb.connect(user='test_user',passwd='123456',db='gic',port=18806,host='192.68.21.22',charset='utf8')
                                         # 连接所测系统的 MySQL 数据库
    cursor = coon.cursor()
    aa=cursor.execute(sql)               # 将执行 SQL 语句
    info = cursor.fetchmany(aa)
```

```python
    for ii in info:
        id_list = []
        id_list.append(ii)
    return str(id_list[0][0])                    # 返回第一个标签的 id 的值
    coon.commit()
    cursor.close()
    coon.close()                                 # 关闭释放数据库连接

def deltagname(param):                           # 获取所要删除 tagname，根据数据库中的 tagname 值作为参数
    sql = "select tagname from t_tag_info where shopid=1000169 and status=1"    #根据店铺id和状态查找标签名称
    coon = MySQLdb.connect(user='test_user',passwd='123456',db='gic',port=18806,host='192.68.21.22',
charset='utf8')
                                                 # 连接所测系统的 MySQL 数据库
    cursor = coon.cursor()
    aa=cursor.execute(sql)                       # 将执行 SQL 语句
    info = cursor.fetchmany(aa)
    for ii in info:
        name_list = []
        name_list.append(ii)
    return name_list[0][0]                       # 返回第一个 tagname 的值
    coon.commit()
    cursor.close()
    coon.close()                                 # 关闭释放数据库连接

def writeResult(case_id,result):                           # 写测试结果到数据库
    result = result.encode('utf-8')                        # 测试结果转换成 UTF-8 格式编码
    now = time.strftime("%Y-%m-%d %H:%M:%S")               # 当前时间格式话为此形式
    sql = "UPDATE zt_testrun set lastRunResult= %s, lastRunDate= %s, lastRunner='auto' where zt_testrun.task=9 and zt_testrun.`case`=%s;"
    param = (result,now,case_id)                           # 把测试结果、时间、用例 id，作为动
                                                           # 态参数写到数据库
    coon = MySQLdb.connect(user='root',passwd='test123456',db='zentao',port=3306,host='192.168.58.79',
charset='utf8')    #连接禅道 MySQL 数据库，host 一定要为 IP 地址
    cursor = coon.cursor()
    cursor.execute(sql,param)                              #执行带参数的 SQL 语句
    coon.commit()
    cursor.close()
    coon.close()                                           #关闭数据库连接
```

```python
def writeBug(bug_id,interface_name,request,response,res_check):    #写测试 bug 到数据库
    interface_name = interface_name.encode('utf-8')                #接口名称字段格式转码为中文
    res_check = res_check.encode('utf-8')                          #校验字段格式转码为中文
    response = response.encode('utf-8')                            #接口响应数据字段格式转码为中文
    request = request.encode('utf-8')                              #接口请求数据字段格式转码为中文
    now = time.strftime("%Y-%m-%d %H:%M:%S")                       #写按此格式的当前时间
    bug_title = str(bug_id) + '_' + interface_name + '_出错了'      #写 bug 的标题格式和内容信息
    step = '[请求报文]'+request+'<br/>'+'[预期结果]'+res_check+'<br/>'+'[响应报文]'+response
    #写 bug 的重现步骤，即出错接口的详情到数据库
    sql = "INSERT INTO `zt_bug` (`openedDate`, `openedBy`, `lastEditedDate`, `lastEditedBy`, `status`,"\
    "`assignedTo`,`assignedDate`, `title`, `keywords`, `hardware`, `openedBuild`, `testtask`, `mailto`,"\
    "`steps`, `storyVersion`, `resolvedDate`, `resolvedBy`, `closedDate`, `closedBy`, `linkBug`, `case`,"\
    "`result`, `module`, `confirmed`, `resolution`, `duplicateBug`, `product`, `activatedCount`, `pri`,"\
    "`severity`) VALUES ('%s', '邹辉', '%s', '1', 'Active', "\
    "'邹辉', '%s', '%s', 'autotest', '1', '1', '1', '系统管理员', "\
    "'%s','1', '%s' , '邹辉', '%s', 'admin', '', '1', "\
    "'1', '1', '1', 'admin', '0', '3', '0', '1', "\
    "'1');"%(now,now,now,bug_title,step,now,now)
#SQL 语句，其中%s 为参数变量：时间，bug_title=bug 标题，step=重现步骤，其他的字段名和值一一对应
    coon = MySQLdb.connect(user='root',passwd='test123456',db='zentao',port=3306,host='192.168.58.79',charset='utf8')
    #连接禅道 MySQL 数据库，host 一定要为 IP 地址
    cursor = coon.cursor()
    cursor.execute(sql)                  # 执行上述 SQL 语句，写 bug 到数据
    coon.commit()
    cursor.close()
    coon.close()                         # 关闭数据库连接

if __name__ == '__main__':
    readSQLcase()                        # 执行 SQL 中的测试用例以及相关的操作 print 'Done!'
```

（4）界面接口测试用例，卖家订单详情流程接口脚本如图 7.3 所示。

▲图 7.3

程序清单 7-5　Python 脚本

python bizOrderDetailV1_2.py　　　　　　　卖家订单详情流程接口脚本如下
# -*- coding:utf-8 -*-
import requests, MySQLdb, time, re
import urllib, urllib2

HOSTNAME = '192.168.215.55' # 公共变量主机 IP

def readSQLcase():　　　　　　　　# 读取测试用例，读取产品编号为 3，用例编号 id 为 158 的流程接口用例
　　sql="SELECT t2.`case`,`title`,t2.`desc`,`precondition`,t2.`expect`,`keywords`,t2.`version` from `zt_case`,`zt_casestep` as t2 inner join (select zt_casestep.`case`, max(version) as tv from `zt_casestep`　group by zt_casestep.`case`) as t1 on t1.`case`=t2.`case` and t1.tv=t2.version where zt_case.id = t2.`case` and zt_case.type='config' and zt_case.product=3 and zt_case.id=158 "
　　coon = MySQLdb.connect(user='root',passwd='test123456',db='zentao',port=3306,host='192.168.58.79',charset='utf8')
　　　　　　　　　　　　　　　　　　# 连接禅道 MySQL 数据库，host 一定要为 IP 地址
　　cursor = coon.cursor()
　　aa=cursor.execute(sql)　　　　　　# 将执行上条 SQL 语句
　　info = cursor.fetchmany(aa)
　　for ii in info:
　　　　case_list = []
　　　　case_list.append(ii)
　　　　interfaceTest(case_list)　　　# 循环读取 SQL 中一条测试用例
　　coon.commit()

```python
        cursor.close()
        coon.close()                            # 关闭和释放数据库连接

def interfaceTest(case_list):                   # 读取一条接口测试用例
    res_flags = []
    request_urls = []
    responses = []
    storenameinfo = re.compile('{storename}')   # 判断参数中如果有动态参数{storename}则在代码中取值，
                                                # 对应作为这个店铺名称
    preorderinfo = re.compile('{preOrderSN}')   # 判断参数中如果有动态参数{preOrderSN}则在代码中取值，
                                                # 对应作为这个预提交订单号
    orderinfo = re.compile('{ordersn}')         # 判断参数中如果有动态参数{ordersn}则在代码中取值，对
                                                # 应作为这个订单编号
    for case in case_list:
        try:
            case_id = case[0]                   # 读取用例的关键信息用例 id
            interface_name = case[1]            # 读取用例的关键信息接口名称对应      标题
            method = case[3]                    # 读取用例的关键信息接口方法对应      方法
            url = case[2]                       # 读取用例的关键信息接口 URL 地址对应  keyword
            param = case[4]                     # 读取用例的关键信息接口参数列表对应    参数
            res_check = case[5]                 # 读取用例的关键信息接口返回校验对应    预期结果
        except Exception,e:
            return '测试用例格式不正确！ %s'%e   #抛异常处理
        if param== '':
            new_url = 'http://'+HOSTNAME+url    #如果接口没有参数，则接口设为 hostname+url
        elif param== 'null':
            new_url = 'http://'+HOSTNAME+url    #如果接口参数为 null,则接口设为 hostname+url。注意，
                                                #当接口没有参数时，可在页面中输入 null 来判断有没
                                                #有参数
        else:
            param = storenameinfo.sub(storename,param)
#如有动态关联参数{storename}，则用值替换变量参数名，即把返回值传给此店铺名称
            param = preorderinfo.sub(preOrderSN,param)
#如有动态关联参数{preOrderSN}，则用值替换变量参数名，即把返回值传给此预提交订单号
            param = orderinfo.sub(ordersn,param)
#如有动态关联参数{ordersn}，则用值替换变量参#数名，即把返回值传给此订单号
            new_url = 'http://'+HOSTNAME+url+'?'+urlParam(param)
#如果接口有参数，则接口设为 hostname+url+param
```

```python
            request_urls.Append(new_url)
        if method.upper() == 'GET':
# 如果为 GET 方法，则读取如下 GET 请求和返回数据
            print str(case_id)+' ' +new_url                          # 打印用例 id 和接口地址
            headers = {'Host':HOSTNAME,
                       'Connection':'keep-alive',
                       'token':token,
                       'Content-Type': 'application/x-www-form-urlencoded',
                       'User-Agent': 'Apache-HttpClient/4.2.6 (Java 1.5)'}
# 设置 HTTP 头信息，包括主机名和用户登录的 token 值等
            data = None
            results = requests.get(new_url,data,headers=headers).text
# 发送 GET 请求，results 得到请求的返回数据
            responses.append(results)
            res = readRes(results,res_check)
# 对请求的返回数据进行校验、正则表达式校验。校验结果有三种：pass、fail、GIGF
            print results
            if 'pass' == res:
                writeResult(case_id,'pass')                          # 写结果为 pass 到这个关联用例 id
                res_flags.append('pass')
            else:
                res_flags.append('fail')
                writeResult(case_id,'fail')                          # 写结果为 fail 到这个关联用例 id
                if reserror(results):
                    writeBug(case_id,interface_name,new_url,"api response is error",res_check)
#如果是接口响应异常，即服务器异常时，这种方式即直接打印出错信息记录 bug 写到数据库
                else:
                    writeBug(case_id,interface_name,new_url,results,res_check)
#如果是接口校验数据错误，则用这种方式即把该接口的请求和响应数据记录 bug 写到数据库
        else:
#如果不为 GET 方法，则读取如下 POST 请求和返回数据
            print str(case_id)+' ' +new_url                          # 打印用例 id 和接口地址
            headers = {'Host':HOSTNAME,
                       'Connection':'keep-alive',
                       'token':token,
                       'Content-Type': 'application/x-www-form-urlencoded',
                       'User-Agent': 'Apache-HttpClient/4.2.6 (Java 1.5)'
                       }
# 设置 HTTP 头信息，包括主机名和用户登录的 token 值
```

```python
            data = None
            results = requests.post(new_url,data,headers=headers).text
# 发送 POST 请求，results 得到请求的返回数据
            responses.append(results)
            res = readRes(results,res_check)
# 对请求的返回数据进行校验、正则表达式校验。校验结果有三种：pass、fail、JIJR
            print results
            if 'pass' == res:
                writeResult(case_id,'pass')                    # 写结果为 pass 到这个关联用例 id
                res_flags.append('pass')
            else:
                res_flags.append('fail')
                writeResult(case_id,'fail')                    # 写结果为 fail 到这个关联用例 id
                if reserror(results):
                    writeBug(case_id,interface_name,new_url,"api response is error",res_check)
#如果是接口响应异常，即服务器异常时，用这种方式即直接打印出错信息记录 bug 并写到数据库
                else:
                    writeBug(case_id,interface_name,new_url,results,res_check)
#如果是接口校验数据错误，则用这种方式即把该接口的请求和响应数据记录 bug 并写到数据库
            try:
                GetToken.token(results)            #获取登录接口后的 token 值，作为动态参数传递
            except:
                print 'ok1'
            try:
                preOrderSN(results)                #获取预提交订单号，作为动态参数传递
            except:
                print 'ok2'
            try:
                ordersn(results)                   #获取订单编号，作为动态参数传递
            except:
                print 'ok3'

def readRes(res,res_check):                        #校验结果，如一致返回 pass，否则返回错误提示
    res = res.replace('":"','=').replace('":','=')  #校验时，将":"替换为=，再进行校验
    res_check = res_check.split(';')
    for s in res_check:
        if s in res:
            pass
        else:
```

```python
            return '错误，返回参数和预期结果不一致'+str(s)
    return 'pass'

def urlParam(param):                                    #参数值的替换
    param1=param.replace('*','&')                       #如果参数在数据库中为*，则替换为&
    return param1.replace('&quot;','\"')
#如果参数在数据库中为&quot;替换成"，则这是因为在页面中输入"时，存储在数据库中就变成了&quot;,
#所以要转换

def token(results):                                     #取登录接口中token参数进行匹配
    global token
    regx = '.*"token":"(.*)","ud"'                      #token的正则表达式，左匹配"token":"，右匹配","ud"
    pm = re.search(regx, results)
    if pm:
        token = pm.group(1).encode('utf-8')             #如果匹配到，则返回token值
        print token
        return True
    return False

def preOrderSN(results):                                #预提交订单号参数取动态值
    global preOrderSN
    regx = '.*"preOrderSN":"(.*)","toHome"'             #预提交订单号取值的正则表达式，左匹配
                                                        #"preOrderSN":" 右匹配"toHome"
    pm = re.search(regx, results)
    if pm:
        preOrderSN = pm.group(1).encode('utf-8')        #如果匹配到，则转换为中文并返回值
        return preOrderSN
    return False

def ordersn(results):                                   #订单编号参数取动态值
    global ordersn
    regx = '.*"tradeNo":"(.*)"}'                        #预提交订单号取值的正则表达式，左匹配"tradeNo":"，右匹配"}
    pm = re.search(regx, results)
    if pm:
        ordersn = pm.group(1).encode('utf-8')           #如果匹配到，则转换为中文并返回值
        return ordersn
    return False
```

```
def reserror(results):
#接口页面返回匹配到的 HTML 页面服务器没有响应，则改调用接口错误
    global html
    regx = 'html'
    pm = re.search(regx, results)
#regx 变量赋值为 HTML 字符串，如果服务器异常时，则会返回 404 等 HTML 标示，进行匹配
    if pm:
        return regx
    return False

def storename(param):              #店铺名称防止重复，设置动态参数，即用 test+时间戳的方式命名店铺名称
    global storename               #定义店铺名称全局变量
    storename='test'+str(time.time())    #设置店铺名称 test+时间戳
    return storename               #返回店铺名称


def writeResult(case_id,result):              #写测试结果到数据库
    result = result.encode('utf-8')           #测试结果转换成 UTF-8 格式编码
    now = time.strftime("%Y-%m-%d %H:%M:%S")     #将当前时间格式化为此形式
    sql = "UPDATE zt_testrun set lastRunResult= %s, lastRunDate= %s, lastRunner='auto' where zt_testrun.task=9 and zt_testrun.`case`=%s;"
    param = (result,now,case_id)              #把测试结果、时间、用例 id，作为动态参数写到数据库
    print result
    coon = MySQLdb.connect(user='root',passwd='test123456',db='zentao',port=3306,host='192.168.58.79',charset='utf8')
    #连接禅道 MySQL 数据库，host 一定要为 IP 地址
    cursor = coon.cursor()
    cursor.execute(sql,param)                 #执行带参数的 SQL 语句
    coon.commit()
    cursor.close()
    coon.close()                              #关闭和释放数据库连接

def writeBug(bug_id,interface_name,request,response,res_check):    #写测试 bug 到数据库
    interface_name = interface_name.encode('utf-8')    #接口名称字段格式转码为中文
    res_check = res_check.encode('utf-8')              #校验字段格式转码为中文
    response = response.encode('utf-8')                #接口响应数据字段格式转码为中文
    request = request.encode('utf-8')                  #接口请求数据字段格式转码为中文
    now = time.strftime("%Y-%m-%d %H:%M:%S")           #写按此格式的当前时间
```

```
        bug_title = str(bug_id) + '_' + interface_name + '_出错了'         #写 bug 的标题格式和内容信息
        step = '[请求报文]<br />'+request+'<br/>'+'[预期结果]<br/>'+res_check+'<br/>'+'<br/>'+'[响应报文]<br />'+'<br/>'+response#写 bug 的重现步骤和出错接口的详情到数据库
        sql = "INSERT INTO `zt_bug` (`openedDate`, `openedBy`, `lastEditedDate`, `lastEditedBy`, `status`,"\
        "`assignedTo`,`assignedDate`, `title`, `keywords`, `hardware`, `openedBuild`, `testtask`, `mailto`,"\
        "`steps`, `storyVersion`, `resolvedDate`, `resolvedBy`, `closedDate`, `closedBy`, `linkBug`, `case`,"\
        "`result`, `module`, `confirmed`, `resolution`, `duplicateBug`, `product`, `activatedCount`, `pri`,"\
        "`severity`) VALUES ('%s', '邹辉', '%s', '1', 'Active', "\
        "'邹辉', '%s', '%s', 'autotest', '1', '1', '1', '系统管理员', "\
        "'%s','1', '%s' , '邹辉', '%s', 'admin', '', '1', "\
        "'1', '1', '1', 'admin', '0', '1', '0', '1', "\
        "'1');"%(now,now,now,bug_title,step,now,now)
#SQL 语句，其中，%s 为参数变量时间，bug_title=bug 标题，step=重现步骤，其他字段名和值一一对应
        coon = MySQLdb.connect(user='root',passwd='test123456',db='zentao',port=3306,host='192.168.58.79',charset='utf8')
                                                # 连接禅道 MySQL 数据库，host 一定要为 IP 地址
        cursor = coon.cursor()
        cursor.execute(sql)                     # 执行上述 SQL 语句，写 bug 信息到数据库
        coon.commit()
        cursor.close()
        coon.close()                            # 关闭和释放数据库连接

class GetToken(object):
    @staticmethod
    def token(results):                         # 把 token 定义成静态方法 进行匹配和传递
        global token
        regx = '.*"token":"(.*)","ud"'          # token 的正则表达式，左匹配"token":" 右匹配","ud"
        pm = re.search(regx, results)
        if pm:
            token = pm.group(1).encode('utf-8') #如果匹配到，则返回 token 值
            return True
        return False

if __name__ == '__main__':
    readSQLcase()                               #执行 SQL 中的测试用例以及相关操作
    print 'Done!'
    time.sleep(60)
```

（5）界面接口测试用例，卖家订单发货流程接口如图 7.4 所示。

132 | 软件自动化测试开发

编号	参数	预期
1	/buyer/user/login.do	phone=13798359580*pwd=47ec2dd791e31e2ef2076caf64ed9b
2	/buyer/cart/updatecart.do	action=3*amount=1*goodsid=20394
3	/buyer/cart/listgoods.do	null
4	/buyer/cart/settle.do	goodsids=20394
5	/buyer/order/submitpreorder.do	receiverinfoid=661*receipt=0*preordersinfo=[{"preOrders":[{"deliverySeq":"1","comment":"","preOrderSN":"{preOrderSN}"}]}]*paytype=1
6	/buyer/order/payorder.do	ordersns=[ordersn]
7	/buyer/order/wxorderseq.do	paychannel=1*tradeno=[ordersn]
8	/buyer/order/orderstatusinform.do	tradeno={ordersn}*status=1*tradeseq=*vk=
9	/portal/user/login.do	phone=13798359580*pwd=47ec2dd791e31e2ef2076caf64ed9b
10	/portal/order/deliverygoods.do	ordersn={ordersn}*orderstatus=40*ordercomment=.

▲图 7.4

程序清单 7-6　Python 脚本

python bizOrderDeliveryV1_2.py　　　　　　　　卖家订单发货流程接口

```python
# -*- coding:utf-8 -*-
import requests, MySQLdb, time, re
import urllib, urllib2

HOSTNAME = '192.168.215.55'        #公共变量主机 IP

def readSQLcase():                  #读取测试用例，读取产品编号为 3,用例编号 id 为 157 的流程接口用例
    sql="SELECT t2.`case`,`title`,t2.`desc`,`precondition`,t2.`expect`,`keywords`,t2.`version` from `zt_case`,`zt_casestep` as t2 inner join (select zt_casestep.`case`, max(version) as tv from `zt_casestep`    group by zt_casestep.`case`) as t1 on t1.`case`=t2.`case` and t1.tv=t2.version where zt_case.id = t2.`case` and zt_case.type='config' and zt_case.product=3 and zt_case.id=157 "
    coon = MySQLdb.connect(user='root',passwd='test123456',db='zentao',port=3306,host='192.168.58.79',charset='utf8')
                                    #连接禅道 MySQL 数据库，host 一定要为 IP 地址
    cursor = coon.cursor()
    aa=cursor.execute(sql)          #将执行上条 SQL 语句
    info = cursor.fetchmany(aa)
    for ii in info:
        case_list = []
        case_list.append(ii)
        interfaceTest(case_list)    # 循环读取 SQL 中一条测试用例
```

```
        coon.commit()
        cursor.close()
        coon.close()                    # 关闭和释放数据库连接

def interfaceTest(case_list):           # 读取一条接口测试用例
    res_flags = []
    request_urls = []
    responses = []
    storenameinfo = re.compile('{storename}')
    # 判断参数中如果有动态参数{storename}，则在代码中取值，对应作为这个店铺名称
    preorderinfo = re.compile('{preOrderSN}')
    # 判断参数中如果有动态参数{preOrderSN}，则在代码中取值，对应作为这个预提交订单号
    orderinfo = re.compile('{ordersn}')
    # 判断参数中如果有动态参数{ordersn}，则在代码中取值，对应作为这个订单编号
    tradeinfo = re.compile('{tradeno}')
    # 判断参数中如果有动态参数{tradeno}，则在代码中取值，对应作为这个交易号
    for case in case_list:
        try:
            case_id = case[0]                   # 读取用例的关键信息用例 id
            interface_name = case[1]            # 读取用例的关键信息接口名称      对应标题
            method = case[3]                    # 读取用例的关键信息接口方法      对应方法
            url = case[2]                       # 读取用例的关键信息接口 URL 地址  对应 keyword
            param = case[4]                     # 读取用例的关键信息接口参数列表   对应参数
            res_check = case[5]                 # 读取用例的关键信息接口返回校验   对应预期结果
        except Exception,e:
            return '测试用例格式不正确！ %s'%e    #抛异常处理
        if param== '':
            new_url = 'http://'+HOSTNAME+url    #如果接口没有参数，则接口设为 hostname+url
        elif param== 'null':
            new_url = 'http://'+HOSTNAME+url    #如果接口参数 null 接口设为 hostname+url。注意，当
                                                #接口没有参数时，可在页面中输入 null 来判断有没有
                                                #参数
        else:
            param = storenameinfo.sub(storename,param)
#如有动态关联参数{storename}， 则用值替换变量参数名，即把返回值传给此店铺名称
            param = preorderinfo.sub(preOrderSN,param)
#如有动态关联参数{preOrderSN}， 则用值替换变量参数名，即把返回值传给此预提交订单号
            param = orderinfo.sub(ordersn,param)
```

```
                #如有动态关联参数{ordersn}，则用值替换变量参数名，即把返回值传给此订单号
                    param = tradeinfo.sub(tradeno,param)
                #如有动态关联参数{tradeno}，则用值替换变量参数名，即把返回值传给此交易号
                    new_url = 'http://'+HOSTNAME+url+'?'+urlParam(param)
                #如果接口有参数，则接口设为 hostname+url+param
                    request_urls.append(new_url)
                    if method.upper() == 'GET':                    #如果为 GET 方法，则读取如下 GET 请求和返回数据
                        print str(case_id)+' '+new_url             #打印用例 id 和接口地址
                        headers = {'Host':HOSTNAME,
                                    'Connection':'keep-alive',
                                    'token':token,
                                    'Content-Type': 'application/x-www-form-urlencoded',
                                    'User-Agent': 'Apache-HttpClient/4.2.6 (Java 1.5)'}
                # 设置 HTTP 头信息，包括主机名和用户登录的 token 值等
                        data = None
                        results = requests.get(new_url,data,headers=headers).text
                # 发送 GET 请求，results 得到请#求的返回数据
                        responses.append(results)
                        res = readRes(results,res_check)
                # 对请求的返回数据进行校验、正则表达式校验。校验结果有三种：pass、 fail、JUF
                        print results
                        if 'pass' == res:
                            writeResult(case_id,'pass')            # 写结果为 pass 到这个关联用例 id
                            modifyOrderStatus(tradeno)             # 如果用例 pass，则设置订单状态变更
                            modifyOrderStatus2(tradeno)            # 如果用例 pass，则设置订单状态变更 2
                            res_flags.append('pass')
                        else:
                            res_flags.append('fail')
                            writeResult(case_id,'fail')            # 写结果为 fail 到这个关联用例 id
                            writeBug(case_id,interface_name,new_url,results,res_check)
                #如果是接口校验数据错误，则用这种方式记录 bug 写到数据库，以及该接口的请求和响应数据
                    else:
                #如果不为 GET 方法，则读取如下 POST 请求和返回数据
                        print str(case_id)+' '+new_url             # 打印用例 id 和接口地址
                        headers = {'Host':HOSTNAME,
                                    'Connection':'keep-alive',
                                    'token':token,
                                    'Content-Type': 'application/x-www-form-urlencoded',
```

```
                        'User-Agent': 'Apache-HttpClient/4.2.6 (Java 1.5)'
                    }                               # 设置 HTTP 头信息，包括主机名和用户登录的 token 值
            data = None
            results = requests.post(new_url,data,headers=headers).text
# 发送 POST 请求，results 得到请求的返回数据
            responses.append(results)
            res = readRes(results,res_check)
# 对请求的返回数据进行校验、正则表达式校验。校验结果有三种 pass、fail、JIJF
            print results
            if 'pass' == res:
                writeResult(case_id,'pass')         # 写结果为 pass 到这个关联用例 id
                res_flags.append('pass')
            else:
                res_flags.append('fail')
                writeResult(case_id,'fail')         # 写结果为 fail 到这个关联用例 id
                if reserror(results):
                    writeBug(case_id,interface_name,new_url,"api response is error",res_check)    #如果是接
#口响应异常，即服务器异常时，可用这种方式记录 bug 并写到数据库，以及该接口的请求和响应数据
                else:
                    writeBug(case_id,interface_name,new_url,results,res_check)
# 如果是接口校验数据错误，则用这种方式记录 bug 写到数据库，以及该接口的请求和响应数据
            try:
                GetToken.token(results)             # 获取登录接口后的 token 值，作为动态参数传递
            except:
                print 'ok1'
            try:
                preOrderSN(results)                 # 获取预提交订单号，作为动态参数传递
            except:
                print 'ok2'
            try:
                ordersn(results)                    # 获取订单编号，作为动态参数传递
            except:
                print 'ok3'
            try:
                tradeno(results)                    # 获取交易编号，作为动态参数传递
            except:
                print 'ok4'

def readRes(res,res_check):                         # 校验结果如一致则返回 pass ，否则返回错误提示
```

```python
            res = res.replace('":"','="').replace('":','="')    # 校验时将":"替换为=，再进行校验
            res_check = res_check.split(';')
            for s in res_check:
                if s in res:
                    pass
                else:
                    return '错误，返回参数和预期结果不一致'+str(s)
            return 'pass'

def urlParam(param):                            #参数值的替换
    param1=param.replace('*','&')               #如果参数在数据库中为*，则替换为&
    return param1.replace('&quot;','\"')
#如果参数在数据库中为&quot;，则替换成"。这是因为在页面中输入"时，存储在数据库中就变成了&quot;，
#所以要转换

def token(results):                             #取登录接口中 token 参数进行匹配
    global token
    regx = '.*"token":"(.*)","ud"'              #token 的正则表达式，左匹配"token":"，右匹配","ud"
    pm = re.search(regx, results)
    if pm:
        token = pm.group(1).encode('utf-8')     #如果匹配到，则返回 token 值
        print token
        return True
    return False


def preOrderSN(results):                        #预提交订单号参数取动态值
    global preOrderSN
    regx = '.*"preOrderSN":"(.*)","toHome"'
#预提交订单号取值的正则表达式，左匹配"preOrderSN":"，右匹配"toHome"
    pm = re.search(regx, results)
    if pm:
        preOrderSN = pm.group(1).encode('utf-8')    #如果匹配到，则转换为中文并返回值
        return preOrderSN
    return False


def ordersn(results):                           #订单编号参数取动态值
    global ordersn
    regx = '.*"tradeNo":"(.*)"}'                #预提交订单号取值的正则表达式，左匹配"tradeNo":"，右匹配"}
    pm = re.search(regx, results)
```

```python
        if pm:
            ordersn = pm.group(1).encode('utf-8')    #如果匹配到，则转换为中文并返回值
            return ordersn
        return False

def tradeno(results):                                #交易号参数取动态值
    global tradeno
    regx = '.*"brief":"(.*)","tradeNo":"2016'        #预提交订单号取值的正则表达式，左匹配"brief":" 右匹配
    #'","tradeNo":"2016'
    pm = re.search(regx, results)
    if pm:
        tradeno = pm.group(1).encode('utf-8')        #如果匹配到则转换为中文并返回值
        return tradeno
    return False

def modifyOrderStatus(tradeno):                      #根据交易号，修改测试系统中数据库的值
    sql = "update t_order set order_status = 20,sync_step=10 where order_sn = %s;"
                                                     #设置 SQL 语句，根据交易号，更新订单状态和同步状态
    param = (tradeno)                                #定义参数为交易号
    coon = MySQLdb.connect(user='test_user',passwd='123456',db='trade',port=18806,host='192.68.21.22',charset='utf8')
                                                     #连接所测系统的数据库
    cursor = coon.cursor()
    cursor.execute(sql,param)                        #执行修改订单状态的 SQL 语句，带参数
    coon.commit()
    cursor.close()
    coon.close()                                     #关闭释放数据库连接

def modifyOrderStatus2(tradeno):                     #根据交易号，修改测试系统中数据库的值
    sql = "update t_order set order_status = 30,sync_step=10 where order_sn = %s;"
    #设置 SQL 语句，根据交易号，更新订单状态和同步状态
    param = (tradeno)
    coon = MySQLdb.connect(user='test_user',passwd='123456',db='trade',port=18806,host='192.68.21.22',charset='utf8')
     #连接所测系统的数据库
    cursor = coon.cursor()
    cursor.execute(sql,param)                        #执行修改订单状态的 sql 语句，带参数
    coon.commit()
    cursor.close()
```

```python
        coon.close()                                    #关闭释放数据库连接

def reserror(results):
#若接口页面返回匹配到的 HTML 页面服务器没有响应，则改为调用接口错误
    global html
    regx = 'html'
    pm = re.search(regx, results)
    #regx 变量赋值为 HTML 字符串，如果服务器异常，则返回 404 等 HTML 标示，进行匹配
    if pm:
        return regx
    return False

def storename(param):          #防止店铺名称重复，设置动态参数，即 test+时间戳的方式命名店铺名称
    global storename           #定义店铺名称全局变量
    storename='test'+str(time.time())                   # 设置店铺名称 test+时间戳
    return storename                                    # 返回店铺名称


def writeResult(case_id,result):                        # 写测试结果到数据库
    result = result.encode('utf-8')                     # 测试结果转换成 UTF-8 格式编码
    now = time.strftime("%Y-%m-%d %H:%M:%S")            # 当前时间格式化为此形式
    sql = "UPDATE zt_testrun set lastRunResult= %s, lastRunDate= %s, lastRunner='auto' where zt_testrun.task=9 and zt_testrun.`case`=%s;"
    param = (result,now,case_id)
    # 把测试结果、时间、用例 id，作为动态参数写到数据库
    print result
    coon = MySQLdb.connect(user='root',passwd='test123456',db='zentao',port=3306,host='192.168.58.79',charset='utf8')
    #连接禅道 MySQL 数据库，host 一定要为 IP 地址
    cursor = coon.cursor()
    cursor.execute(sql,param)                           # 执行带参数的 SQL 语句
    coon.commit()
    cursor.close()
    coon.close()                                        # 关闭和释放数据库连接

def writeBug(bug_id,interface_name,request,response,res_check):    # 写测试 bug 到数据库
    interface_name = interface_name.encode('utf-8')     # 接口名称字段格式转码为中文
    res_check = res_check.encode('utf-8')               # 校验字段格式转码为中文
```

```python
        response = response.encode('utf-8')                          # 接口响应数据字段格式转码为中文
        request = request.encode('utf-8')                            # 接口请求数据字段格式转码为中文
        now = time.strftime("%Y-%m-%d %H:%M:%S")                     # 写按此格式的当前时间
        bug_title = str(bug_id) + '_' + interface_name + '_出错了'    # 写 bug 的标题格式和内容信息
        step = '[请求报文]<br />'+request+'<br/>'+'[预期结果]<br/>'+res_check+'<br/>'+'<br/>'+'[响应报文]<br
/>'+'<br/>'+response                                                 #写 bug 的重现步骤即出错接口的详情到数据库
        sql = "INSERT INTO `zt_bug` (`openedDate`, `openedBy`, `lastEditedDate`, `lastEditedBy`, `status`,"\
"`assignedTo`,`assignedDate`, `title`, `keywords`, `hardware`, `openedBuild`, `testtask`, `mailto`,"\
"`steps`, `storyVersion`, `resolvedDate`, `resolvedBy`, `closedDate`, `closedBy`, `linkBug`, `case`,"\
"`result`, `module`, `confirmed`, `resolution`, `duplicateBug`, `product`, `activatedCount`, `pri`,"\
"`severity`) VALUES ('%s', '邹辉', '%s', '1', 'Active', "\
"'邹辉', '%s', '%s', 'autotest', '1', '1', '1', '系统管理员', "\
"'%s','1', '%s', '邹辉', '%s', 'admin', '', '1', "\
"'1', '1', '1', 'admin', '0', '1', '0', '1', "\
"'1');"%(now,now,now,bug_title,step,now,now)                         #SQL 语句，其中, %s 为参数变量时间, #bug_title=bug
                                                                     #标题，step=重现步骤，其他字段名和值一一对应
        coon = MySQLdb.connect(user='root',passwd='test123456',db='zentao',port=3306,host='192.168.58.79',
charset='utf8')
#连接禅道 MySQL 数据库, host 一定要为 IP 地址
        cursor = coon.cursor()
        cursor.execute(sql)                                          #执行上述 SQL 语句，写 bug 到数据库
        coon.commit()
        cursor.close()
        coon.close()                                                 #关闭和释放数据库连接


class GetToken(object):
    @staticmethod
    def token(results):                                              #把 token 定义成静态方法，进行匹配和传递
        global token
        regx = '.*"token":"(.*)","ud"'                               #token 的正则表达式，左匹配"token":"，右匹配","ud"
        pm = re.search(regx, results)
        if pm:
            token = pm.group(1).encode('utf-8')                      #如果匹配到，则返回 token 值
            return True
        return False
```

```
if __name__ == '__main__':
    readSQLcase()                        #执行 SQL 中的测试用例及相关操作
    print 'Done!'
    time.sleep(60)
```

程序清单 7-7　　Python 脚本

（6）python report.py，买家和卖家接口测试报告脚本如下：

```
# -*- coding:utf-8 -*-
import requests, MySQLdb, time, re, sys
import urllib, urllib2
from random import choice

HOSTNAME = '192.168.215.55'

def buyer_readAPIbug():                  #读取当天的接口 bug
    sql="SELECT `title`,`steps` from `zt_bug`  where zt_bug.product = 1 and date(openedDate) = curdate(); "   #产品 1 的今天相关 bug 的 SQL 语句
    coon = MySQLdb.connect(user='root',passwd='test123456',db='zentao',port=3306,host='192.168.58.79',charset='utf8')           #连接数据库，host 一定要用真实 IP
    cursor = coon.cursor()
    aa=cursor.execute(sql)               #执行 SQL 语句
    info = cursor.fetchmany(aa)
    for ii in info:                      #循环打印所有 bug
        bug_list = []
        bug_list.append(ii)
        print buyercaseid(bug_list[0][0])+' '+bug_list[0][1]    #打印出 bug 标题和重现步骤
    coon.commit()
    cursor.close()
    coon.close()                         #关闭数据库连接


def buyercaseid(results):                #读取测试用例的接口 id
    global buyerid
    global buyertitle
    regx = r'(\d+)_'                     #正则表达式数字匹配 id
    regx2= u"([\u4e00-\u9fa5]+)"         #正则表达式文字匹配标题
    pm = re.search(regx, results)
    pm2= re.search(regx2,results)
```

```
    if pm:
        buyerid = pm.group(1).encode('utf-8')        #正则表达式返回 true，赋值给 id
    if pm2:
        buyertitle = pm2.group(1)                    #正则表达式返回 true，赋值给标题
    return buyerid+' '+buyertitle                    #返回字符串格式的用例 id 和标题

def buyer_Interface_total():                         #读取接口总数
    sql = "select count(*) from zt_case ,`zt_testrun`,`zt_testtask` where zt_case.type='interface' and zt_case.product=1 and zt_case.id=zt_testrun.`case` and zt_testrun.task=zt_testtask.id and zt_testtask.id=8 and zt_case.status='normal';"
    coon = MySQLdb.connect(user='root',passwd='test123456',db='zentao',port=3306,host='192.168.58.79',charset='utf8')
    cursor = coon.cursor()
    aa=cursor.execute(sql)
    info = cursor.fetchmany(aa)
    return info[0][0]
    coon.commit()
    cursor.close()
    coon.close()


def buyer_case_total():                              #读取测试用例总数
    sql = "select count(*) from zt_case ,`zt_testrun`,`zt_testtask` where zt_case.type='interface' and zt_case.product=1 and zt_case.id=zt_testrun.`case` and zt_testrun.task=zt_testtask.id and zt_testtask.id=8 and zt_case.status='normal';"
    coon = MySQLdb.connect(user='root',passwd='test123456',db='zentao',port=3306,host='192.168.58.79',charset='utf8')
    cursor = coon.cursor()
    aa=cursor.execute(sql)
    info = cursor.fetchmany(aa)
    return info[0][0]
    coon.commit()
    cursor.close()
    coon.close()


def buyer_case_pass():                               #读取通过测试用例总数
    sql = "select count(*) from zt_case ,`zt_testrun`,`zt_testtask` where zt_case.type='interface' and zt_case.product=1 and zt_case.id=zt_testrun.`case` and zt_testrun.task=zt_testtask.id and zt_testtask.id=8 and zt_case.status='normal' and zt_testrun.lastRunResult='pass';"
    coon = MySQLdb.connect(user='root',passwd='test123456',db='zentao',port=3306,host='192.168.58.79',charset='utf8')
    cursor = coon.cursor()
```

```python
        aa=cursor.execute(sql)
        info = cursor.fetchmany(aa)
        return info[0][0]
        coon.commit()
        cursor.close()
        coon.close()

def buyer_case_skip():                    #读取跳过未执行的测试用例总数
        skipcase=biz_case_total()-biz_case_pass()-biz_case_fail()
        return skipcase

def buyer_case_fail():                    #读取失败测试用例总数
        sql = "select count(*) from zt_case ,`zt_testrun`,`zt_testtask` where zt_case.type='interface' and zt_case.product=1 and zt_case.id=zt_testrun.`case` and zt_testrun.task=zt_testtask.id and zt_testtask.id=8 and zt_case.status='normal' and zt_testrun.lastRunResult='fail';"
        coon = MySQLdb.connect(user='root',passwd='test123456',db='zentao',port=3306,host='192.168.58.79',charset='utf8')
        cursor = coon.cursor()
        aa=cursor.execute(sql)
        info = cursor.fetchmany(aa)
        return info[0][0]
        coon.commit()
        cursor.close()
        coon.close()

def buyer_flow_total():                   #读取流程类测试用例总数
        sql = "select count(*) from zt_case ,`zt_testrun`,`zt_testtask` where zt_case.type='config' and zt_case.product=1 and zt_case.id=zt_testrun.`case` and zt_testrun.task=zt_testtask.id and zt_testtask.id=8 and zt_case.status='normal';"
        coon = MySQLdb.connect(user='root',passwd='test123456',db='zentao',port=3306,host='192.168.58.79',charset='utf8')
        cursor = coon.cursor()
        aa=cursor.execute(sql)
        info = cursor.fetchmany(aa)
        return info[0][0]
        coon.commit()
        cursor.close()
        coon.close()
```

```python
def buyer_flow_pass():                          #读取流程类通过测试用例总数
    sql = "select count(*) from zt_case ,`zt_testrun`,`zt_testtask` where zt_case.type='config' and zt_case.product=1 and zt_case.id=zt_testrun.`case` and zt_testrun.task=zt_testtask.id and zt_testtask.id=8 and zt_case.status='normal' and zt_testrun.lastRunResult='pass';"
    coon = MySQLdb.connect(user='root',passwd='test123456',db='zentao',port=3306,host='192.168.58.79',charset='utf8')
    cursor = coon.cursor()
    aa=cursor.execute(sql)
    info = cursor.fetchmany(aa)
    return info[0][0]
    coon.commit()
    cursor.close()
    coon.close()


def buyer_flow_skip():                          #读取流程类跳过测试用例总数
    skipcase=biz_flow_total()-biz_flow_pass()-biz_flow_fail()
    return skipcase


def buyer_flow_fail():                          #读取流程类失败用例总数
    sql = "select count(*) from zt_case ,`zt_testrun`,`zt_testtask` where zt_case.type='config' and zt_case.product=1 and zt_case.id=zt_testrun.`case` and zt_testrun.task=zt_testtask.id and zt_testtask.id=8 and zt_case.status='normal' and zt_testrun.lastRunResult='fail';"
    coon = MySQLdb.connect(user='root',passwd='test123456',db='zentao',port=3306,host='192.168.58.79',charset='utf8')
    cursor = coon.cursor()
    aa=cursor.execute(sql)
    info = cursor.fetchmany(aa)
    return info[0][0]
    coon.commit()
    cursor.close()
    coon.close()


def biz_readAPIbug():                           #读取卖家的 bug
    sql="SELECT `title`,`steps` from `zt_bug`  where zt_bug.product = 3 and date(openedDate) = curdate(); "
    coon = MySQLdb.connect(user='root',passwd='test123456',db='zentao',port=3306,host='192.168.58.79',charset='utf8')
    cursor = coon.cursor()
    aa=cursor.execute(sql)
    info = cursor.fetchmany(aa)
    for ii in info:
```

```python
        bug_list = []
        bug_list.append(ii)
        print bizcaseid(bug_list[0][0])+' '+bug_list[0][1]
    coon.commit()
    cursor.close()
    coon.close()


def bizcaseid(results):                    #正则匹配用例 id 和标题
    global bizid
    global biztitle
    regx = r'(\d+)_'
    regx2= u"([\u4e00-\u9fa5]+)"
    pm = re.search(regx, results)
    pm2= re.search(regx2, results)
    if pm:
        bizid = pm.group(1).encode('utf-8')
    if pm2:
        biztitle= pm2.group(1)#.encode('utf-8')
    return bizid+' '+biztitle


def biz_interface_total():                 #读取卖家接口总数
    sql = "select count(*) from zt_case ,`zt_testrun`,`zt_testtask` where zt_case.type='interface' and zt_case.product=3 and zt_case.id=zt_testrun.`case` and zt_testrun.task=zt_testtask.id and zt_testtask.id=9 and zt_case.status='normal';"
    coon = MySQLdb.connect(user='root',passwd='test123456',db='zentao',port=3306,host='192.168.58.79',charset='utf8')
    cursor = coon.cursor()
    aa=cursor.execute(sql)
    info = cursor.fetchmany(aa)
    return info[0][0]
    coon.commit()
    cursor.close()
    coon.close()


def biz_case_total():                      #读取卖家用例总数
    sql = "select count(*) from zt_case ,`zt_testrun`,`zt_testtask` where zt_case.type='interface' and zt_case.product=3 and zt_case.id=zt_testrun.`case` and zt_testrun.task=zt_testtask.id and zt_testtask.id=9 and zt_case.status='normal';"
    coon = MySQLdb.connect(user='root',passwd='test123456',db='zentao',port=3306,host='192.168.58.79',charset='utf8')
```

```python
    cursor = coon.cursor()
    aa=cursor.execute(sql)
    info = cursor.fetchmany(aa)
    return info[0][0]
    coon.commit()
    cursor.close()
    coon.close()

def biz_case_pass():                    #读取通过用例的总数
    sql = "select count(*) from zt_case ,`zt_testrun`,`zt_testtask` where zt_case.type='interface' and zt_case.product=3 and zt_case.id=zt_testrun.`case` and zt_testrun.task=zt_testtask.id and zt_testtask.id=9 and zt_case.status='normal' and zt_testrun.lastRunResult='pass';"
    coon = MySQLdb.connect(user='root',passwd='test123456',db='zentao',port=3306,host='192.168.58.79',charset='utf8')
    cursor = coon.cursor()
    aa=cursor.execute(sql)
    info = cursor.fetchmany(aa)
    return info[0][0]
    coon.commit()
    cursor.close()
    coon.close()

def biz_case_skip():                    #读取跳过未执行用例总数
    skipcase=biz_case_total()-biz_case_pass()-biz_case_fail()
    return skipcase

def biz_case_fail():                    #读取失败用例总数
    sql = "select count(*) from zt_case ,`zt_testrun`,`zt_testtask` where zt_case.type='interface' and zt_case.product=3 and zt_case.id=zt_testrun.`case` and zt_testrun.task=zt_testtask.id and zt_testtask.id=9 and zt_case.status='normal' and zt_testrun.lastRunResult='fail';"
    coon = MySQLdb.connect(user='root',passwd='test123456',db='zentao',port=3306,host='192.168.58.79',charset='utf8')
    cursor = coon.cursor()
    aa=cursor.execute(sql)
    info = cursor.fetchmany(aa)
    return info[0][0]
    coon.commit()
    cursor.close()
    coon.close()

def biz_flow_total():                   #读取卖家流程用例总数
```

```python
    sql = "select count(*) from zt_case ,`zt_testrun`,`zt_testtask` where zt_case.type='config' and zt_case.product=3 and zt_case.id=zt_testrun.`case` and zt_testrun.task=zt_testtask.id and zt_testtask.id=9 and zt_case.status='normal';"
    coon = MySQLdb.connect(user='root',passwd='test123456',db='zentao',port=3306,host='192.168.58.79',charset='utf8')
    cursor = coon.cursor()
    aa=cursor.execute(sql)
    info = cursor.fetchmany(aa)
    return info[0][0]
    coon.commit()
    cursor.close()
    coon.close()

def biz_flow_pass():                    #读取卖家流程用例通过总数
    sql = "select count(*) from zt_case ,`zt_testrun`,`zt_testtask` where zt_case.type='config' and zt_case.product=3 and zt_case.id=zt_testrun.`case` and zt_testrun.task=zt_testtask.id and zt_testtask.id=9 and zt_case.status='normal' and zt_testrun.lastRunResult='pass';"
    coon = MySQLdb.connect(user='root',passwd='test123456',db='zentao',port=3306,host='192.168.58.79',charset='utf8')
    cursor = coon.cursor()
    aa=cursor.execute(sql)
    info = cursor.fetchmany(aa)
    return info[0][0]
    coon.commit()
    cursor.close()
    coon.close()

def biz_flow_skip():                    #读取卖家流程用例未执行总数
    skipcase=biz_flow_total()-biz_flow_pass()-biz_flow_fail()
    return skipcase

def biz_flow_fail():                    #读取卖家流程用例失败总数
    sql = "select count(*) from zt_case ,`zt_testrun`,`zt_testtask` where zt_case.type='config' and zt_case.product=3 and zt_case.id=zt_testrun.`case` and zt_testrun.task=zt_testtask.id and zt_testtask.id=9 and zt_case.status='normal' and zt_testrun.lastRunResult='fail';"
    coon = MySQLdb.connect(user='root',passwd='test123456',db='zentao',port=3306,host='192.168.58.79',charset='utf8')
    cursor = coon.cursor()
    aa=cursor.execute(sql)
    info = cursor.fetchmany(aa)
    return info[0][0]
    coon.commit()
```

```
        cursor.close()
        coon.close()

if __name__ == '__main__':
    print u'接口自动化汇总测试报告'
    print '--------------------'+u'买家'+'--------------------'
    print u'接口总数'+'    '+str(buyer_Interface_total())
    print u'独立用例总数'+'    '+str(buyer_case_total())+'    '+u'通过数'+'    '+str(buyer_case_pass())+'    '+u'跳过数'+'    '+str(buyer_case_skip())+'    '+u'失败数'+'    '+str(buyer_case_fail())
    print u'流程用例总数'+'    '+str(buyer_flow_total())+'    '+u'通过数'+'    '+str(buyer_flow_pass())+'    '+u'跳过数'+'    '+str(buyer_flow_skip())+'    '+u'失败数'+'    '+str(buyer_flow_fail())
    print
    print '--------------'+u'今天运行的接口错误详情如下'+'--------------'
    buyer_readAPIbug()

    print
    print

    print '--------------------'+u'卖家'+'--------------------'
    print u'接口总数'+'    '+str(biz_interface_total())
    print u'独立用例总数'+'    '+str(biz_case_total())+'    '+u'通过数'+'    '+str(biz_case_pass())+'    '+u'跳过数'+'    '+str(biz_case_skip())+'    '+u'失败数'+'    '+str(biz_case_fail())
    print u'流程用例总数'+'    '+str(biz_flow_total())+'    '+u'通过数'+'    '+str(biz_flow_pass())+'    '+u'跳过数'+'    '+str(biz_flow_skip())+'    '+u'失败数'+'    '+str(biz_flow_fail())
    print
    print '--------------'+u'今天运行的接口错误详情如下'+'--------------'
    biz_readAPIbug()
    print 'Done!'
```

7.2 Python 接口测试数据展示

1. 接口 bug 的清单如图 7.5 所示。

▲图 7.5

2．接口 bug 的详情如图 7.6 所示。

▲图 7.6

3．接口测试用例列表如图 7.7 所示。

▲图 7.7

4．接口异常测试用例清单如图 7.8 所示。

▲图 7.8

5．单个接口测试用例详情如图 7.9 所示。

▲图 7.9

6．流程类接口测试用例如图 7.10 所示。

▲图 7.10

7. 版本测试清单如图 7.11 所示。

图 7.11

8. 接口版本用例测试的结果有三种：通过 or 失败 or 未执行，如图 7.12 所示

第 7 章 API 接口自动化源代码 | 151

图 7.12

7.3 脚本持续集成到 Jenkins

只需把脚本放到 Jenkns 取代码的 SVN 上，在 Jenkins 上加入运行脚本的命令即可，用 7.4 节最前面的脚本执行如下命令：

python testAPIbuyerV1_2.py
python OrderSubmitFlowV1_2.py
python OrderCancelFlowV1_2.py
python testAPIbizV1_2.py
python bizOrderDetailV1_2.py
python bizOrderDeliveryV1_2.py
python report.py

7.4 接口自动化测试报告

全部通过时的报告如下（后续改进方案——可写成 HTML 格式，生成更详细直观的报表）。

1. 接口自动化汇总报告 示例 1

--------------------买家--------------------

接口总数　　76

独立用例总数　76　　通过数　76　　跳过数　0　　失败数　0

流程用例总数　12　　通过数　12　　跳过数　0　　失败数　0

---------------今天运行的接口错误详情如下---------------

--------------------卖家--------------------

接口总数　63

独立用例总数　63　　通过数　63　　跳过数　0　　失败数　0

流程用例总数　2　　通过数　2　　跳过数　0　　失败数　0

---------------今天运行的接口错误详情如下---------------

Done!

2．接口自动化汇总测试报告　示例2

--------------------买家--------------------

接口总数　76

独立用例总数　76　　通过数　74　　跳过数　0　　失败数　2

流程用例总数　12　　通过数　12　　跳过数　0　　失败数　0

---------------今天运行的接口错误详情如下---------------

194　[接 口 名 称]http://192.168.215.55/test1/updatecart.do?action=1&amount=1&goodsid=20000000460000001019&specid=2000000046000000101900004&storeid=46
[预期结果]state=0
[响应报文]{"msg":"商品已下架","state":3822002}

196　[接 口 名 称]http://192.168.215.55/test1/settle.do?goodslist=&storeid=46&receiverinfoid=851
[预期结果]state=0
[响应报文]{"msg":"结算失败，商品已经下架或库存不足","data":{"goodsStatusResult":0,"goodsStatusList":[{"specId":"2000000046000000101900004","goodsId":"20000000460000001019","status":"1

"}]},"state":3829003}

--------------------卖家--------------------

接口总数　　60

独立用例总数　　60　　通过数　　59　　跳过数　　0　　失败数　　1

流程用例总数　　2　　通过数　　2　　跳过数　　0　　失败数　　0

--------------今天运行的接口错误详情如下--------------

222[接口名称]http://192.168.215.55/test2/upgoods.do?optype=3&productlist=3945
[预期结果]state=0
[响应报文]{"msg":"系统繁忙，请稍后再试","state":3720001}

Done!

第 8 章

Selenium 的 Web 自动化测试

8.1 Selenium 自动化测试准备

1. Selenium 介绍

Selenium 是一个 Web 开源自动化测试框架，页面级操作，模拟用户真实操作，API 从系统层面触发事件。

Selenium 1.0

Sever/Client 工作方式，可在 local 或 remote 机器上运行基于 js 注入的 case 底层。

为什么一定要用代理服务器的模式？答案是同源策略，它是由 Netscape 提出的一个著名的安全策略，现在所有可支持 Javascript 的浏览器都在使用这个策略。

Selenium 2.0

Selenium 2.0 基于 Selenium 1.0（即 Javascript ）并结合其 WebDriver 来模拟用户的真实操作。WebDriver 原生绑定到浏览器，绕过浏览器安全模型。它有很好的处理 Ajax 的能力，并且支持多种浏览器（如 Safari、IE、Firefox、Chrome 等），可以运行在多种操作系统上面。目前，大家几乎都在使用 Selenium 2.0。

2. 基于 Java 开发

（1）Selenium IDE 录制用例，回放，导出 Java 代码。

（2）多种方式定位并控制页面元素：

- Web 元素定位
- id name linktext xpath tag css
- 异常处理

（3）自动化测试用例封装和设计原则参考如图 8.1 所示。

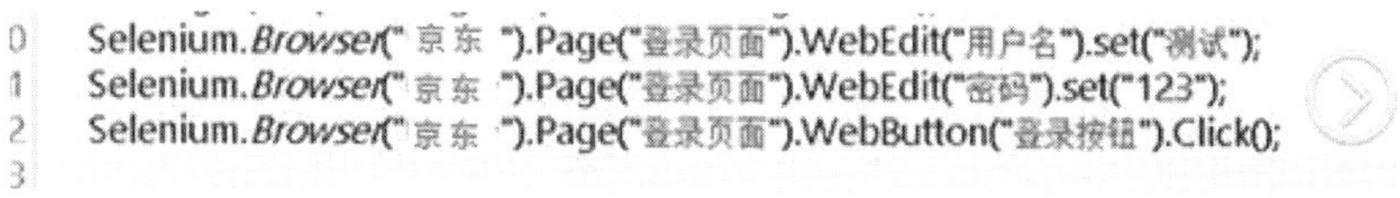

```
0    Selenium.Browser(" 京东 ").Page("登录页面").WebEdit("用户名").set("测试");
1    Selenium.Browser(" 京东 ").Page("登录页面").WebEdit("密码").set("123");
2    Selenium.Browser(" 京东 ").Page("登录页面").WebButton("登录按钮").Click();
3
```

▲图 8.1

- 安装 JDK，配置 JDK 环境变量。
- 安装 Eclipes。

3．运行时注意驱动与浏览器的版本

默认为火狐（Firefox）驱动，如 chromedriver.exe、IEDriverServer.exe。

> 注意，版本需要兼容 JDK 1.7 以及 IE 8 到 IE 10, IE 11 及以上需要更新驱动才能支持，请大家自行研究更新版本驱动或采用旧版本浏览器。

4．环境搭建的简要步骤

（1）安装 JDK，配置 JDK 环境变量。

（2）安装 Eclipes。

（3）安装火狐 Selenium IDE、谷歌 Chrome drive 和 IE driver。

（4）加入 jxl、log4j、Selenium 包。

（5）加入 JUnit 包，创建 JUnit 测试类。

（6）加入 TestNG 包，创建 TestNG 测试类 TestNG.xml。

（7）安装 Ant，配置 Ant 环境变量 build.xml。

（8）安装 Jenkins、Tomcat，配置 Tomcat 环境变量。

（9）节点 Slave 的配置及连接。

（10）启动节点自动化测试。

8.2　Selenium 自动化源码解析

登录 demo 的工程文件路径：https://pan.baidu.com/s/1i4UwtkL

密码：3xfb

（基于 Java 语言和 Selenium 自动化框架工具开发）

```
shopping
---src                                          --Java 类
-----logo
-----logo.log                                   --日志类
-------------- SelLogger
-------------- log4j.properties
------logo.module                               --模块公共类
-------------- BrowserDriver
-------------- FileExcel
-------------- Login
-------------- PublicModule
-------------- SiteUrl
------logo.testsuite.login                      --登录测试类
-------------- LoginTestSuiteJunit
-------------- LoginTestSuitetestng
------testng.xm                                 --运行测试用例配置文件
JUnit4                                          --JUnit4 组件包
TestNG                                          --TestNG 组件包
ant                                             --Ant 组件包
classes                                         --编译类
lib                                             --环境依赖包
logs                                            --日志生成文件
test-output                                     --测试结果报告
tomcat                                          --Jenkins 运行环境
build.xml                                       --Ant 构建运行配置文件
jxl.jar                                         -Excel 组件包
selenium-java-2.43.1-srcs.jar                   --Selenium 组件包
selenium-java-2.43.1.jar                        --Selenium 组件包
selenium-server-standalone-2.39.0.jar           --Selenium 组件包
test.xls                                        --参数数据 Excel 文件
```

如图 8.2 所示。

▲图 8.2

下面一起来看看写 Web 测试自动化框架的源代码。

程序清单 8-1　Java 代码

SelLogger.java

package logo.log;

import java.io.FileInputStream;
import java.io.IOException;
import java.io.InputStream;
import java.text.SimpleDateFormat;
import java.util.Calendar;
import java.util.Properties;
import org.apache.log4j.Logger;
import org.apache.log4j.PropertyConfigurator;
import org.testng.Reporter;

```java
public class SelLogger {
    private static Logger logger = null;
    private static SelLogger logg = null;
    public static SelLogger getLogger(Class<?> T) {
        if (logger == null) {
            Properties props = new Properties();
            try {
                InputStream is = new FileInputStream("src//logo//log//log4j.properties");
//根据 log4 级今日志配置文件读取日志信息
                props.load(is);
            } catch (IOException e) {
                e.printStackTrace();
            }
            PropertyConfigurator.configure(props);
            logger = Logger.getLogger(T);
            logg = new SelLogger();
        }
        return logg;
    }

    //重写 logger 方法
    public void log(String msg) {
        SimpleDateFormat sdf = new SimpleDateFormat("yyyy-MM-dd HH:mm:ss"); //日志中日期格式化
        Calendar ca = Calendar.getInstance();
```

```
            logger.info(msg);
            Reporter.log("Reporter:" + sdf.format(ca.getTime()) + "===>" + msg);
    }

    public void debug(String msg) {                                      //debug 级别日志
        logger.debug(msg);
    }

    public void warn(String msg) {                                       //wran 级别日志
        logger.warn(msg);
        Reporter.log("Reporter:" + msg);
    }

    public void error(String msg) {                                      //error 级别日志
        logger.error(msg);
        Reporter.log("Reporter:" + msg);
    }
}
```

程序清单 8-2 Java 代码

log4j.properties

log4j.rootLogger=Info, stdout, logfile

log4j.appender.stdout=org.apache.log4j.Consoleappappender
log4j.appender.stdout.layout=org.apache.log4j.PatternLayout
log4j.appender.stdout.layout.ConversionPattern=%-d{yyyy-MM-dd HH:mm:ss,SSS} %p %t [%c]%M(line:%L)%m%n

log4j.appender.logfile.encoding=UTF-8
log4j.appender.logfile=org.apache.log4j.DailyRollingFileappappender
log4j.appender.logfile.File=logs/shopping.log
log4j.appender.logfile.layout=org.apache.log4j.PatternLayout
log4j.appender.logfile.layout.ConversionPattern=%-d{yyyy-MM-dd HH:mm:ss,SSS} %p %t %M(line:%L)%m%n

BrowserDriver.java

package logo.module;

import org.openqa.selenium.WebDriver;
import org.openqa.selenium.firefox.FirefoxDriver;

```java
import org.openqa.selenium.ie.InternetExplorerDriver;
import org.openqa.selenium.remote.DesiredCapabilities;

public class BrowserDriver                                                      //浏览器驱动的类
{
    private WebDriver browser;                                                  //创建浏览器对象

    public WebDriver browser()
    {
        browser = new FirefoxDriver();                                          //浏览器对象赋值给 Firefox 浏览器的驱动
        return browser;
    }

    public WebDriver browser1()
    {
      System.setProperty("webdriver.ie.driver", ".\\lib\\IEDriverServer.exe");  //浏览器对象赋值给 IE 浏览器的
                                                                                //驱动
      DesiredCapabilities capabilities = DesiredCapabilities.internetExplorer();
capabilities.setCapability(InternetExplorerDriver.INTRODUCE_FLAKINESS_BY_IGNORING_SECURITY_DOMAINS, true);
      browser = new InternetExplorerDriver(capabilities);
      return browser;
    }

    public WebDriver browser2()
    {
      System.setProperty("webdriver.ie.driver", ".\\lib\\chromedriver.exe");    //浏览器对象赋值给谷歌浏
                                                                                //览器的驱动
      DesiredCapabilities capabilities = DesiredCapabilities.internetExplorer();
capabilities.setCapability(InternetExplorerDriver.INTRODUCE_FLAKINFSS_BY_IGNORING_SECURITY_DOMAINS, true);
      browser = new InternetExplorerDriver(capabilities);
      return browser;
    }
}
```

程序清单 8-3　Java 代码

FileExcel.java

```java
package logo.module;

import java.io.File;
import java.io.IOException;
import jxl.Cell;
import jxl.Sheet;
import jxl.Workbook;
import jxl.read.biff.BiffException;
import logo.log.SelLogger;

public class FileExcel
{
    private static final SelLogger logger = SelLogger.getLogger(FileExcel.class);

    public String username()                           //根据文件获取用户名的类
    {
       try
       {
            String fileName = "./test.xls";            //Excel 文件相对路径，用户名参数化文件数据
            File file = new File(fileName);
            Workbook wb = Workbook.getWorkbook(file);  //获取文件
            Sheet sheet = wb.getSheet(0);              //获取文件的 sheet
            Cell cell = sheet.getCell(0, 0);           //第 2 个单元格
            String username = cell.getContents();      //获取一个单元格的值返回给用户名
            return username;
        } catch (BiffException e)                      //异常处理
        {
            e.printStackTrace();
        } catch (IOException e)
        {
            e.printStackTrace();
        }
        return null;
    }
```

```java
    public String password()                    //根据文件获取密码的类
    {
        try
        {
            String fileName = "./test.xls";      //Excel 文件相对路径，密码参数化文件数据
            File file = new File(fileName);
            Workbook wb = Workbook.getWorkbook(file);   //获取文件
            Sheet sheet = wb.getSheet(0);        //获取文件的 sheet
            Cell cell = sheet.getCell(1, 0);     //第 2 个单元格
            String password = cell.getContents();   //获取第 2 个单元格的值返回给密码
            return password;
        } catch (BiffException e)                //异常处理
        {
            e.printStackTrace();
        } catch (IOException e)
        {
            e.printStackTrace();
        }
        return null;
    }
}
```

程序清单 8-4　Java 代码

Login.java
package logo.module;

import logo.log.SelLogger;
import org.junit.Assert;
import org.openqa.selenium.WebDriver;
import org.openqa.selenium.WebElement;
import org.openqa.selenium.support.FindBy;
import org.openqa.selenium.support.How;
import org.openqa.selenium.support.PageFactory;
import org.openqa.selenium.support.pagefactory.AjaxElementLocatorFactory;
import org.openqa.selenium.support.pagefactory.ElementLocatorFactory;

public class Login
{
    private static final SelLogger logger = SelLogger.getLogger(Login.class);

```java
private WebDriver driver;
PublicModule p=new PublicModule();

@FindBy(how = How.ID, using = "loginname")
public static WebElement loginnameInputbox;         //定义一个根据 id 判断的元素赋值为"loginname"

@FindBy(how = How.ID, using = "loginpwd")
public static WebElement loginpwdInputbox;          //定义一个根据 id 判断的元素赋值为"loginpwd"

@FindBy(how = How.ID, using = "btn-login")
public static WebElement loginBtn;                  //定义一个根据 id 判断的元素赋值为"btn-login"

@FindBy(how = How.LINK_TEXT, using = "退出")
public static WebElement loginResult;               //定义一个根据 linkText 判断的元素赋值为"退出"

public Login(WebDriver driver)
{
    this.driver = driver;
    ElementLocatorFactory finder = new AjaxElementLocatorFactory(driver,120);
    PageFactory.initElements(finder, this);
    driver.manage().window().maximize();            //浏览器页面默认为最大化
}

public String userName(String userNameTxt)
{
    loginnameInputbox.clear();                      //清空文本框默认值
    loginnameInputbox.sendKeys(userNameTxt);        //输入用户名
    return userNameTxt;
}

public String passWord(String userPwdTxt)
{
    loginpwdInputbox.clear();                       //清空文本框默认值
    loginpwdInputbox.sendKeys(userPwdTxt);          //输入密码
    return userPwdTxt;
}

public void clickLoginButton()
{
```

```
                p.waitForPageLoadByID("btn-login", driver);      //根据元素 id 等待登录按钮，如果超时则截屏
                loginBtn.click();                                //单击登录按钮
    }

    public void checkResult()
    {
            Assert.assertEquals(loginResult.isDisplayed(), true);
//结果断言，根据 link_text 判断是否存在"退出"，存在则说明登录成功，否则用例失败
    }
}
```

程序清单 8-5　Java 代码

```
PublicModule.java
package logo.module;

import java.io.File;
import java.text.SimpleDateFormat;
import java.util.Date;
import java.util.Set;
import logo.log.SelLogger;
import org.openqa.selenium.By;
import org.openqa.selenium.NoSuchWindowException;
import org.openqa.selenium.OutputType;
import org.openqa.selenium.TakesScreenshot;
import org.openqa.selenium.WebDriver;
import org.openqa.selenium.WebElement;
import org.openqa.selenium.support.ui.ExpectedCondition;
import org.openqa.selenium.support.ui.WebDriverWait;

public class PublicModule
{
    private static final SelLogger logger = SelLogger.getLogger(PublicModule.class);

    public void waitForPageLoadByID(final String ID, WebDriver browser)  //根据 ID 等待页面元素
    {
        try
        {
            WebDriverWait wait = new WebDriverWait(browser, 10);     //浏览器超时等待时间设为 10 秒
            wait.until(new ExpectedCondition<WebElement>()
            {
```

```java
                    public WebElement Apply(WebDriver d)
                    {
                            return d.findElement(By.id(ID));
                    }
                });
                if (null == wait)
                {
                        this.CaptureScreenshot(Thread.currentThread().getId() + "ID",browser);    //超时时自动
                                                                                                  //截屏保存图片
                        browser.quit();                                                           //退出浏览器
                }

        } catch (Exception e)
        {
            e.printStackTrace();
            this.CaptureScreenshot(Thread.currentThread().getId() + "ID", browser);    //未发现页面时自
                                                                                       //动截屏保存图片
            browser.quit();
        }

    }

    public  void waitForPageLoadBylinkText(final String ID, WebDriver browser)         //根据 linkText 等待
                                                                                       //页面元素
{
        try
        {
          WebDriverWait wait = new WebDriverWait(browser,10);                          //浏览器超时等待
                                                                                       //时间设为 10 秒

            wait.until(new ExpectedCondition<WebElement>(){
                    public WebElement apply(WebDriver d) {
                            return d.findElement(By.linkText(ID));
                }});
                if (null == wait)
                    {
                            this.CaptureScreenshot(Thread.currentThread().getId() + "ID",browser);
//超时时自动截屏保存图片
                            browser.quit();
                    }
        } catch (Exception e)
```

```java
            {
                e.printStackTrace();
                this.CaptureScreenshot(Thread.currentThread().getId() + "ID", browser);
//未发现页面时自动截屏保存图片
                browser.quit();
            }
        }

        public boolean switchToWindow_Title(WebDriver driver, String windowTitle) {
//根据 Title 切换新窗口
            boolean flag = false;
            try {
                String currentHandle = driver.getWindowHandle();           //获取当前窗口句柄
                Set<String> handles = driver.getWindowHandles();
                for (String s : handles) {
                    if (s.equals(currentHandle))
                        continue;
                    else {
                        driver.switchTo().window(s);
                        if (driver.getTitle().contains(windowTitle)) {     //根据新窗口的标题 Title 判断，如果
                                                                            //存在则标志为 true，切换成功
                            flag = true;
                            break;
                        } else
                            continue;
                    }
                }
            } catch (NoSuchWindowException e) {
                System.out.println("Window: " + windowTitle + " cound not find!!!"
//如果不存在新窗口的标题，则打印异常，标志为 false，没有找到则改窗口
                        + e.fillInStackTrace());
                flag = false;
            }
            return flag;
        }

        public boolean switchToWindow_Url(WebDriver driver, String windowUrl) {    //根据 URL 切换新窗口
            boolean flag = false;
            try {
                String currentHandle = driver.getWindowHandle();                    //获取当前窗口句柄
```

```java
            Set<String> handles = driver.getWindowHandles();
            for (String s : handles) {
                if (s.equals(currentHandle))
                    continue;
                else {
                    driver.switchTo().window(s);
                    if (driver.getCurrentUrl().contains(windowUrl)) {   //根据新窗口的 URL 判断，如果存
                                                                       //在则标志为 true，切换成功

                        flag = true;
                        break;
                    } else
                        continue;
                }
            }
        } catch (NoSuchWindowException e) {
            System.out.println("Window: " + windowUrl + " cound not find!!!"   //如果不存在新窗口的 URL，则
                                                                               //打印异常，标志为 false，并
                                                                               //提示没有找到这个窗口
                    + e.fillInStackTrace());
            flag = false;
        }
        return flag;
    }

    public void CaptureScreenshot(String fileName, WebDriver driver)           //截屏方法
    {
        String dirName = "test-output/screenshot";                             //截屏保存的文件目录
        if (!(new File(dirName).isDirectory()))
        {
            new File(dirName).mkdir();
        }
        SimpleDateFormat sdf = new SimpleDateFormat("yyyyMMdd-HHmmss");   //时间格式化
        String time = sdf.format(new Date());
        TakesScreenshot tsDriver = (TakesScreenshot) driver;
        File image = new File(dirName + File.separator + time + "_" + fileName + ".png");   //截屏保存的图片
                                                                                            //名称
        tsDriver.getScreenshotAs(OutputType.FILE).renameTo(image);
    }
}
```

程序清单 8-6　Java 代码

SiteUrl.java
package logo.module;

```java
public class SiteUrl
{
    private String indexUrl;                              //定义首页 URL 私有变量
    private String memberUrl;                             //定义用户页 URL 私有变量
    private String itemUrl;                               //定义单品页 URL 私有变量

    public String indexUrl()                              //首页
    {
        indexUrl="http://www.aolaigo.com";                //首页 URL
        return indexUrl;
    }

    public String memberUrl()                             //用户登录页
    {
        memberUrl="http://member.aolaigo.com/login.aspx"; //用户登录页 URL
        return memberUrl;
    }

    public String itemUrl()                               //单品页
    {
        itemUrl="http://item.aolaigo.com";                //单品页 URL
        return itemUrl;
    }

}
```

程序清单 8-7　Java 代码

LoginTestSuiteJunit.java

```java
package logo.testsuite.login;

import static org.junit.Assert.*;
import java.util.concurrent.TimeUnit;
import logo.log.SelLogger;
import logo.module.BrowserDriver;
import logo.module.FileExcel;
import logo.module.Login;
import logo.module.PublicModule;
import logo.module.SiteUrl;
import org.junit.After;
import org.junit.AfterClass;
import org.junit.Before;
import org.junit.BeforeClass;
import org.junit.Test;
import org.openqa.selenium.By;
import org.openqa.selenium.WebDriver;
import org.openqa.selenium.WebElement;
import org.openqa.selenium.firefox.FirefoxDriver;
import org.openqa.selenium.support.ui.ExpectedCondition;
import org.openqa.selenium.support.ui.WebDriverWait;
import org.testng.annotations.Parameters;
import org.tesgng.log4testng.Logger;

public class LoginTestSuiteJunit
{
    private static final SelLogger logger = SelLogger.getLogger(LoginTestSuiteJunit.class);
    private WebDriver browser;                                      //创建浏览器对象
    private String indexUrl,memberUrl,baseitemURL;                  //创建 URL 字符对象
    private boolean acceptNextAlert = true;
    private StringBuffer verificationErrors = new StringBuffer();
    FileExcel f=new FileExcel();                                    //创建 Excel 文件操作对象并初始化
    PublicModule p=new PublicModule();                              //创建公共类对象并初始化
    BrowserDriver browserDriver =new BrowserDriver();               //创建浏览器驱动对象并初始化
    SiteUrl siteUrl=new SiteUrl();                                  //创建 URL 对象并初始化
```

```java
@BeforeClass
public static void setUpBeforeClass() throws Exception
{
}

@AfterClass
public static void tearDownAfterClass() throws Exception
{
}

@Before
public void setUp() throws Exception
{
    browser = browserDriver.browser1();              //初始化，启动 IE 浏览器
    indexUrl =   siteUrl.indexUrl();                 //赋值默认输入的首页 URL
    memberUrl = siteUrl.memberUrl();                 //赋值默认输入的用户页 URL
    baseitemURL = siteUrl.itemUrl();                 //赋值默认输入的单品页 URL
    browser.manage().timeouts().implicitlyWait(30, TimeUnit.SECONDS);
                                                     //浏览器超时等待为 30 秒
}

@After
public void tearDown() throws Exception
{
    browser.quit();                                  //清理环境，并退出关闭浏览器
    String verificationErrorString = verificationErrors.toString();
    if (!"".equals(verificationErrorString))
    {
        fail(verificationErrorString);
    }
}

@Test
public void Login_TestCase_01()                      //登录用例
{
    logger.log("login start...");                    //打印开始登录日志
    browser.get(memberUrl);                          //启动浏览器并输入默认 URL
    Login login = new Login(browser);                //创建登录的类调用登录类初始化
```

```
        login.userName(f.username());          //输入用户名
        login.passWord(f.password());          //输入密码方法
         login.clickLoginButton();             //单击登录按钮
        try {                                  //异常处理
            Thread.sleep(2000);
        } catch (InterruptedException e) {
            e.printStackTrace();
        }
        login.checkResult();                   //校验是否登录成功
        logger.log("...login end"+"\r\n");     //打印结束登录日志
    }
}
```

程序清单 8-8　Java 代码

```
LoginTestSuiteTestNG.java
package logo.testsuite.login;

import logo.log.SelLogger;
import org.testng.annotations.Test;

public class LoginTestSuiteTestNG
{
    private static final SelLogger logger = SelLogger.getLogger(LoginTestSuiteTestNG.class);    //加载 JUnit 类的日
                                                                                                //志信息

    private LoginTestSuiteJunit loginTestsuite = new LoginTestSuiteJunit(); //创建登录 JUnit 的类并初始化

    @Test(groups = {"login"})
    public void Login_phone_TestCase_01() throws Exception     //手机登录用例
    {
        loginTestsuite.setUp();                                //登录用例初始化
        loginTestsuite.Login_TestCase_01();                    //登录用例测试
        loginTestsuite.tearDown();                             //登录用例清理环境
    }
}
```

程序清单 8-9　Java 代码

```xml
testing.xml
<?xml version="1.0" encoding="UTF-8"?>
<suite name="Suite" parallel="false">
  <test name="Test">
    <groups>
      <run>
            <include name="login" />
            <include name="search" />
            <exclude name="order"/>
      </run>
    </groups>

    <classes>
        <class name="logo.testsuite.login.LoginTestSuiteTestNG"/>
            <methods preserve-order="true">
                <include name="Login_phone_TestCase_01" />
            </methods>
        </classes>
  </test> <!-- Test -->
</suite> <!-- Suite -->
```

8.3　持续集成到 Jenkins

将源代码结合 TestNG 和 Ant，持续集成到 Jenkins。下面是 build.xml 文件的内容。

程序清单 8-10　Java 代码

```xml
<project name="shopping" basedir="." default="run_tests">     <!-- 默认调用 run_tests 任务 -->
    <property name="src" value="src" />
    <property name="dest" value="classes" />
    <property name="lib.dir" value="${basedir}/lib" />
    <property name="output.dir" value="${basedir}/test-output" />    <!-- 设置报告输出的路径 -->
```

```xml
<path id="compile.path">                                    <!-- 编译路径设置 -->
    <fileset dir="${lib.dir}/">
        <include name="*.jar" />
    </fileset>
    <pathelement location="${src}" />
    <pathelement location="${dest}" />
</path>

<target name="init">                                        <!-- 初始化设置 -->
    <mkdir dir="${dest}" />
</target>

<target name="compile" depends="init">                      <!-- 编译和初始化 -->
    <echo>compile tests</echo>
    <javac srcdir="${src}" destdir="${dest}" encoding="UTF-8"
        classpathref="compile.path" />
</target>
<taskdef resource="testngtasks" classpath="${lib.dir}/testng.jar" />   <!-- testng.jar 目录和文件 -->

<target name="run_tests" depends="compile">                 <!-- 开始测试 -->
    <echo>running tests</echo>
    <testng classpathref="compile.path" outputdir="${output.dir}"
        haltonfailure="no"
        failureproperty="failed"
        parallel="true"
        threadCount="1" >
        <xmlfileset dir="${src}/" includes="testng.xml" />
<!--结合 testng，调用 testng.xml 里面配置的测试用例  -->
        <classfileset dir="${dest}">
            <include name="/*.class" />
        </classfileset>

    </testng>
    <antcall target="transform" />
    <!-- <fail message="TEST FAILURE" if="failed" /> -->

</target>

<target name="transform" description="report">              <!-- 生成报告 -->
    <xslt
```

```
                in="${output.dir}/testng-results.xml"
                style="${lib.dir}/testng-results.xsl"      <!-- 以 testnt-results.xsl 模板的方式 -->
                out="${output.dir}/Report.html"            <!-- 输出 HTML 格式的测试报告 -->
                force="yes">
                <!-- you need to specify the directory here again -->
                <param name="testngXslt.outputDir" expression="${output.dir}" />
                <classpath refid="compile.path" />
            </xslt>
        </target>
</project>
```

8.4 Web 自动化测试结果展示

3 个用例的测试结果 TestNG 报告如图 8.3 所示。

▲图 8.3

3 个用例，即 1 个手机登录和 2 个搜索用例执行记录的 TestNG 的详细报告如图 8.4 和 8.5 所示。

Test	# Passed	# Skipped	# Failed	Time (ms)	Included Groups	Excluded Groups
Suite						
Test	2	0	1	402,292	search, login	order

Class	Method	Start	Time (ms)
Suite			
Test — failed			
aolaigo.testsuite.login.LoginTestSuiteTestng	Login_phone_TestCase_01	1420789939063	312607
Test — passed			
aolaigo.testsuite.search.SearchTestSuiteTestng	Search_input_TestCase_01	1420790251673	48583
	Search_url_TestCase_02	1420790300257	41081

Test

aolaigo.testsuite.login.LoginTestSuiteTestng#Login_phone_TestCase_01

Reporter:2015-01-09 15:52:42===>login start...

```
org.openqa.selenium.NoSuchElementException: Timed out after 120 seconds. Unable to locate the element
For documentation on this error, please visit: http://seleniumhq.org/exceptions/no_such_element.html
Build info: version: '2.43.1', revision: '5163bce', time: '2014-09-10 16:27:58'
System info: host: 'WIN-1J0MS8955GV', ip: '192.168.115.1', os.name: 'Windows Server 2012', os.arch: 'amd64', os.version:
Driver info: driver.version: unknown
        at org.openqa.selenium.support.pagefactory.AjaxElementLocator.findElement(AjaxElementLocator.java:72)
        at org.openqa.selenium.support.pagefactory.internal.LocatingElementHandler.invoke(LocatingElementHandler.java:37)
        at com.sun.proxy.$Proxy8.isDisplayed(Unknown Source)
        at aolaigo.module.Login.checkResult(Unknown Source)
        at aolaigo.testsuite.login.LoginTestSuiteJunit.Login_TestCase_01(Unknown Source)
        at aolaigo.testsuite.login.LoginTestSuiteTestng.Login_phone_TestCase_01(Unknown Source)
Caused by: org.openqa.selenium.NoSuchElementException: Unable to locate element: {"method":"id","selector":"box1"}
```

▲图 8.4

aolaigo.testsuite.search.SearchTestSuiteTestng#Search_input_TestCase_01

Messages
Reporter:2015-01-09 15:57:44===>search input start...
Reporter:2015-01-09 15:58:18===>...search input end

back to summary

aolaigo.testsuite.search.SearchTestSuiteTestng#Search_url_TestCase_02

Messages
Reporter:2015-01-09 15:58:31===>search url start...
Reporter:2015-01-09 15:58:59===>...search url end

back to summary

▲图 8.5

第 9 章

JMeter 接口测试和性能测试

本章介绍 JMeter 的入门知识。

9.1 安装和介绍

JMeter 安装文件路径：https://pan.baidu.com/s/1kVJdnuv。

JMeter 是轻量级的开源且稳定的自动化测试工具。

思路：在接口说明文档中整理出接口测试案例，其中需要包括详细的入参和出参数据，以及明确的格式和检查点，做到接口用例 100%覆盖，并和开发人员一起对接口测试案例进行评审。

9.1.1 安装 JDK 并配置环境变量

安装 JDK，配置 JDK 环境变量。在系统 cmd 命令提示符下输入 java -version，如果能够看到 Java 版本信息，说明 JDK 安装成功。解压缩 Apache-jmeter-2.12.zip 到 E 盘目录下 E:\apiauto\apache-jmeter-2.12，如图 9.1 所示。

▲图 9.1

设置 JMeter 环境变量，运行 JMeter，直接打开 E:\apache-jmeter-2.12\bin\jmeter.bat 就可以了。

9.1.2 JMeter 主要组件介绍

1．测试计划

测试计划是使用 JMeter 进行测试的开始，是其他 JMeter 测试元件的容器。

给测试计划取一个有含义的项目名称。如图 9.2 所示。

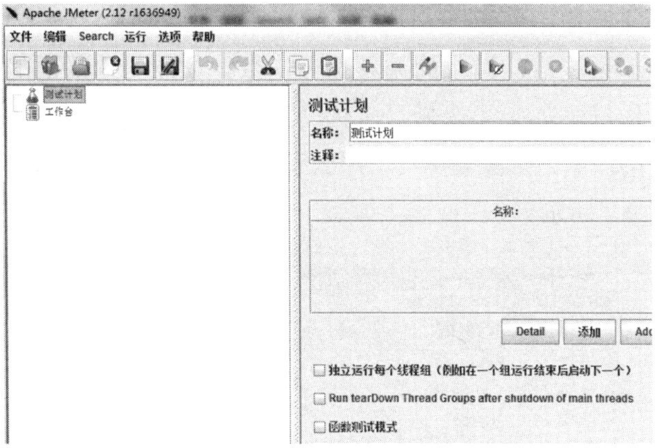

▲图 9.2

2．用户定义的变量

用户可以自己定义变量，在用到这个变量时直接用${变量名}引用。例如，变量名＝orderId，值＝1234567890，在需要使用 1234567890 时直接引用${orderId}即可，如图 9.3 所示。

▲图 9.3

3．线程组

名称：为线程组设置一个有含义的名称。

线程属性－线程数：设置发送请求的用户数，即并发数。

线程属性－Ramp-Up Period（in seconds）：线程间的时间间隔，单位是秒，即所有线程在多少时间内启动。

线程属性－循环次数：请求的循环执行次数。如果勾选后面的"永远"选项，那么请求将一直继续。如果不选择"永远"选项，而是在输入框中输入数字，那么请求将重复指定的次数。如果输入 1，那么请求执行一次；如果输入 0，则请求实际上不执行。如图 9.4 所示。

▲图 9.4

4．取样器（HTTP 请求）

名称：为 HTTP 请求设置一个有含义的名称。

Web 服务器：服务器名称或 IP、端口号，这些会在脚本录制时自动添加，也可以使用"用户自定义变量"。

HTTP 请求：一般是 HTTP 协议，GET 方法或 POST 方法。

同请求一起发送参数：请求中的参数、值可以在这里设置，需要用到参数化或动态数据关联。

同请求一起发送文件：可以制定同请求一起发送哪个文件，比如图片等。如图 9.5 所示。

▲图 9.5

5. 监听器

监听器负责收集测试结果，并给出结果显示方式。常用的包括：查看结果树和聚合报告，两者都支持将结果数据写入文件，其他的可填上去看看就行。聚合报告和查看结果树的截图分别如图 9.6 和图 9.7 所示。

▲图 9.6

▲图 9.7

6．控制器

逻辑控制器可以自定义 JMeter 发送请求的行为逻辑，它与采样器 Sampler 结合使用，可以模拟复杂的请求序列，如图 9.8 所示。

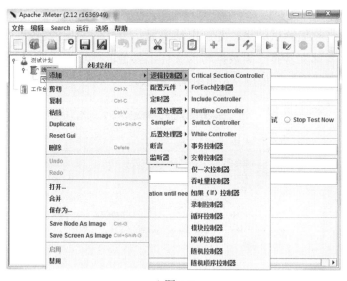

▲图 9.8

循环控制器可以设置请求的循环次数或永远循环（如勾选"永远"选项的话），如图 9.9 所示。

▲图 9.9

事务控制器可将多个请求放在同一个事务中。如果选中 Gegerate parent sample，则聚合报告中只显示事务控制器的数据，而不会显示其中的各个请求的数据，反之则全部显示，如图 9.10 所示。

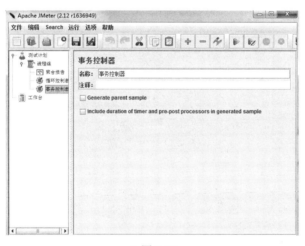

▲图 9.10

7．断言

断言可用来判断请求响应的结果是否如用户所预期的，即在确保功能正确的前提下执行接口或压力测试，因此断言对于有效的测试是很有必要的，如图 9.11 所示。

▲图 9.11

8. 配置元件

配置元件，即维护采样器 Sampler 需要的配置信息，并根据实际需要修改请求的内容。主要是在参数化中用到 CSV Data Set Config，如图 9.12 所示。

▲图 9.12

9. 前置处理器

前置处理器负责在生成请求之前完成工作，常用来修改请求的设置，如图 9.13 所示。

▲图 9.13

10. 后置处理器

后置处理器负责在生成请求之后完成工作，常用来处理响应的数据，主要是在动态关联中用到后置处理器的正则表达式提取器，如图 9.14 所示。

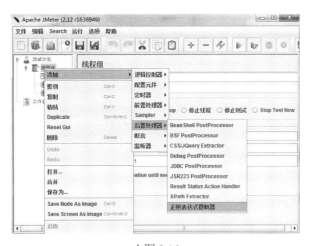

▲图 9.14

11. 定时器

定时器负责定义请求之间的延迟间隔，如固定定时器，如图 9.15 所示。

▲图 9.15

12. 参数化设置

在测试计划中使用的都是 HTTP 请求,在 HTTP 请求中通常会提交一些参数和值,为了避免在每一次请求中都使用相同的参数值,这时就需要进行参数化设置。下面简要介绍两种参数化设置的方法:

- 一种是利用函数助手中的_Random 函数进行参数化设置;
- 另一种是利用配置元件中的 CSV Data Set Config 进行参数化设置。

(1) 函数助手。

JMeter 中有一个函数助手的功能,里面内置了多个函数,可以利用其中的_Random 随机函数来进行请求中的参数化设置。函数助手对话框如图 9.16 所示。

▲图 9.16

（2）CSV Data Set Config。

在线程组上单击右键，在弹出的快捷菜单中选择"添加→配置元件→CSV Data Set Config"命令，如图 9.17 所示。

Filename 文件名称：参数化要引用的文件名，同时在 JMeter 存放 Extras 路径下新建一个 account.txt 文件，并存放相应参数值的数据内容。

File encoding（文件编码）：可不填。

Variable Names（comma-delimited）变量名（用逗号分隔）：多个变量可以引用同一个文件，用逗号分隔。这里是 username,password。

Delimiter（use'\t'for tab）：参数文件中多个变量值的分隔符，\t 表示用 tab 键分隔，默认是逗号。

Recycle on EOF?：结束后是否循环？默认是 True。

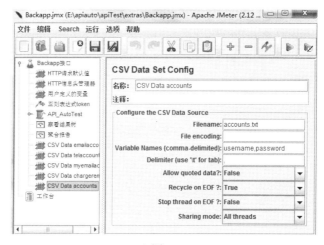

▲图 9.17

account.txt 文件内容如图 9.18 所示。电脑文件路径：E:\apiauto\apache-jmeter-2.12\extras。

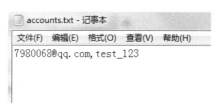

▲图 9.18

13．动态数据关联

在 HTTP 请求的参数中会遇到一些参数的值是从服务器响应返回的动态数据，这些数据需要进行关联才能使得下一次请求能成功地被服务器接受。在 JMeter 中，采用正则表达式提取器来获取这些动态数据。

正则表达式允许用户从服务器响应中获取数据，作为一个后置处理器，正则表达式提取器会在每一个请求执行后再执行。正则表达式提取请求的参数值，产生模板字符串，并将结果保存到设置的相应变量中。

比如要想获得从服务器相应登录请求后的 token 值，就可以使用图 9.19 所示的正则表达式提取器。正则表达式的组成是：左匹配、（取值标志）、右匹配。

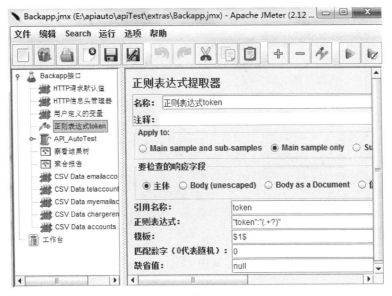

▲图 9.19

使用时可以使用${token}来表示获取到的 token 值，如图 9.20 所示。

▲图 9.20

也可以在 HTTP 头信息中，一起发送下一次的 HTTP 请求，如图 9.21 所示。

▲图 9.21

9.2 Jmeter 接口测试示例

如果是 Web，则需要使用 badboy 等先进行录制，这里讲的是接口，因此可不用录制。

（1）新建测试计划 Backapp，如图 9.22 所示。

▲图 9.22

（2）添加 HTTP 请求默认值，输入服务器名称或 IP、端口号以及中文编码 UTF-8，如图 9.23 所示。

▲图 9.23

(3)添加 HTTP 信息头管理器,输入名称 token 和值${token},如图 9.24 所示。

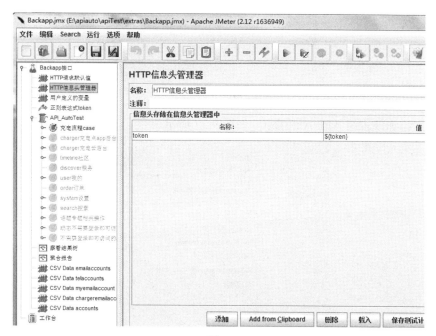

▲图 9.24

（4）添加 Token 的正则表达式："token": "（+?）"，如图 9.25 所示。

▲图 9.25

（5）添加用户自定义变量，输入参数 orderId 和值${orderId}，如图 9.26 所示。

▲图 9.26

（6）添加线程组 API_AutoTest，如图 9.27 所示。

▲图 9.27

（7）添加简单控制器充电流程 case，如图 9.28 所示。

▲图 9.28

（8）添加 HTTP 请求、充电流程登录接口和充电流程电桩，开启充电接口，如图 9.29 所示。

▲图 9.29

（9）设置 HTTP 请求路径/login、参数名 email 和值${username}、参数名 password 和值${password}，如图 9.30 所示。

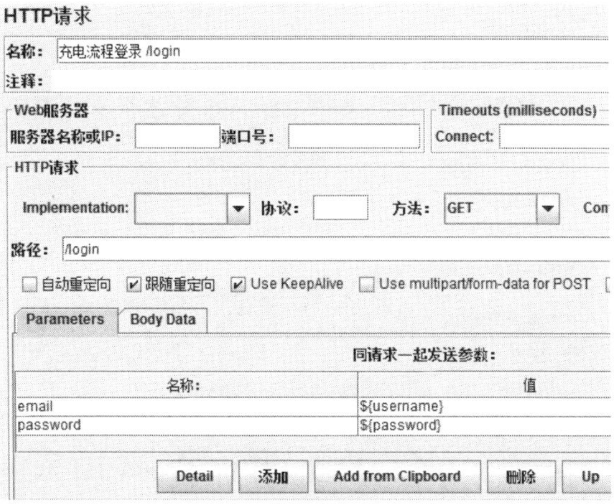

▲图 9.30

（10）输入路径/api/startCharge 、参数名 chargerid 和值 80000111、参数名 plugid 和值 jfS……，如图 9.31 所示。

▲图 9.31

（11）添加响应断言设置，如果包含字符串"code":0，则响应数据正确，如图 9.32 所示。

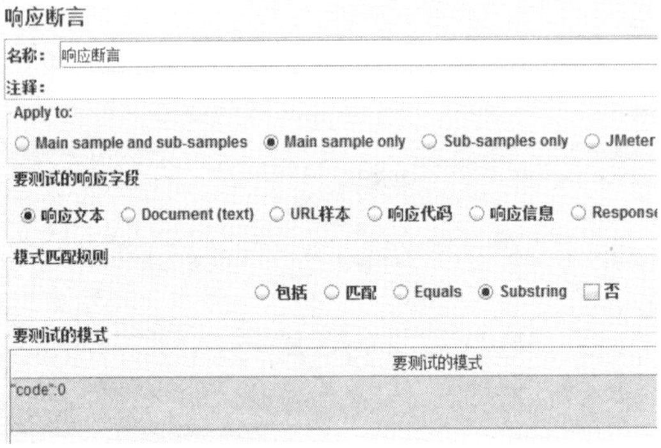

▲图 9.32

（12）添加 CSV Data Set Config，设置文件名 account.txt 以及参数名 username 和 password。

在 JMeter 的 Extras 路径下新建文件 Account.txt 中输入用户名和密码：
7980068@qq.com,test_123。如图 9.33 所示。

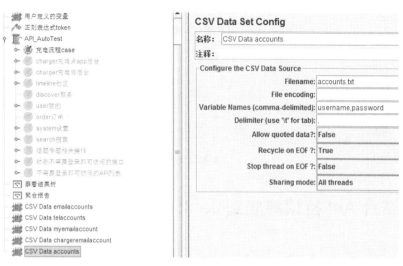

▲图 9.33

（13）添加查看结果树，单击运行按钮，查看响应数据，如图 9.34 所示。

▲图 9.34

（14）添加聚合报告，单击运行按钮，查看响应时间等，如图 9.35 所示。

▲图 9.35

9.3 结合 Ant 持续集成到 Jenkins

把 Ant 及 9.4 节中的接口测试结果报告结合起来。xml 文件中一一对应配置文件名、文件路径和各字段。

Build.xml：

```xml
<?xml version="1.0"?>

<project name="ant-jmeter" default="all">
    <description>
    </description>

    <property name="testpath" value="${user.dir}"/> <!--构建路径设置-->
      <property name="resultpath" value="/Users/zouhui/Documents/workspace"/>
    <property name="jmeter.home" value="${basedir}/.."/>
    <property name="report.title" value="接口测试结果："/>

    <!-- Name of test (without .jmx) -->
    <property name="test" value="Backapp"/>

    <!-- Should report include response data for failures? -->
    <property name="show-data" value="n"/>

    <property name="format" value="2.1"/>
```

```xml
<condition property="style_version" value="">
    <equals arg1="${format}" arg2="2.0"/>
</condition>

<condition property="style_version" value="_21">
    <equals arg1="${format}" arg2="2.1"/>
</condition>

<condition property="funcMode">
    <equals arg1="${show-data}" arg2="y"/>
</condition>

<condition property="funcMode" value="false">
  <not>
    <equals arg1="${show-data}" arg2="y"/>
  </not>
</condition>

<!-- Allow jar to be picked up locally -->
<path id="jmeter.classpath">            <!--编译路径设置-->
    <fileset dir="${basedir}/extras">
       <include name="ant-jmeter*.jar"/>
    </fileset>
</path>

<taskdef
    name="jmeter"
    classpathref="jmeter.classpath"
    classname="org.programmerplanet.ant.taskdefs.jmeter.JMeterTask"/>

<target name="all" depends="run,report"/>

<target name="run">           <!--运行测试设置-->
    <echo>funcMode = ${funcMode}</echo>
    <delete file="${testpath}/extras/${test}.html"/>
    <delete file="${testpath}/extras/${test}.jtl"/>
    <jmeter
        jmeterhome="${jmeter.home}/apiTest"
        testplan ="${testpath}/extras/${test}.jmx"
        resultlog="${testpath}/extras/${test}.jtl">
    <!--
```

```xml
            <jvmarg value="-Xincgc"/>
            <jvmarg value="-Xmx128m"/>
            <jvmarg value="-Dproperty=value"/>
            <jmeterarg value="-qextra.properties"/>
        -->
            <!-- Force suitable defaults -->
            <property name="jmeter.save.saveservice.output_format" value="xml"/>
            <property name="jmeter.save.saveservice.assertion_results" value="all"/>
            <property name="jmeter.save.saveservice.bytes" value="true"/>
            <property name="file_format.testlog" value="${format}"/>
            <property name="jmeter.save.saveservice.response_data.on_error" value="${funcMode}"/>
        </jmeter>
    </target>

    <property name="lib.dir" value="${jmeter.home}/apiTest/lib"/>

    <!-- Use xalan copy from JMeter lib directory to ensure consistent processing with Java 1.4+ -->
    <path id="xslt.classpath">
        <fileset dir="${lib.dir}" includes="xalan*.jar"/>
        <fileset dir="${lib.dir}" includes="serializer*.jar"/>
    </path>
                                        <!--测试报告 -->
    <target name="report" depends="xslt-report,copy-images">
        <echo>Report generated at ${report.datestamp}</echo>
    </target>

    <target name="xslt-report" depends="_message_xalan">
        <tstamp><format property="report.datestamp" pattern="yyyy/MM/dd HH:mm"/></tstamp>
        <xslt
            classpathref="xslt.classpath"
            force="true"
            in="${testpath}/extras/${test}.jtl"
            out="${testpath}/extras/${test}.html"
            style="${basedir}/extras/jmeter-results-detail-report${style_version}.xsl">
<!--测试报告模板-->
            <param name="showData" expression="${show-data}"/>
            <param name="titleReport" expression="${report.title}"/>
            <param name="dateReport" expression="${report.datestamp}"/>
        </xslt>
    </target>
```

```xml
<!-- Copy report images if needed -->
<target name="copy-images" depends="verify-images" unless="samepath">
    <copy file="${basedir}/extras/expand.png" tofile="${testpath}/extras/expand.png"/>
    <copy file="${basedir}/extras/collapse.png" tofile="${testpath}/extras/collapse.png"/>
</target>

<target name="verify-images">
    <condition property="samepath">
            <equals arg1="${testpath}/extras" arg2="${basedir}/extras" />
    </condition>
</target>

<!-- Check that the xalan libraries are present -->
<condition property="xalan.present">
        <and>
               <!-- No need to check all jars; just check a few -->
            <available classpathref="xslt.classpath" classname="org.apache.xalan.processor.TransformerFactoryImpl"/>
            <available classpathref="xslt.classpath" classname="org.apache.xml.serializer.ExtendedContentHandler"/>
        </and>
</condition>

<target name="_message_xalan" unless="xalan.present">
        <echo>Cannot find all xalan and/or serialiser jars</echo>
        <echo>The XSLT formatting may not work correctly.</echo>
        <echo>Check you have xalan and serializer jars in ${lib.dir}</echo>
</target>

</project>
```

9.4 接口测试结果

接口测试结果如图 9.36 所示。

接口测试记录(部分)：

概况

# 样例总数	失败数	成功通过率	响应速度(ms)
55	0	100.00%	447 ms

详情

接口API	# 样例次数	失败数	成功通过率	响应速度(ms)
邮箱用户登录1 /login?email=1234568@qq.com&password=111111	1	0	100.00%	297 ms
我的预约 /book/my?uid=11986b2f-b666-4e7f-b451-e3bd0d411a37	1	0	100.00%	101 ms
手机用户登录2 /login?email=13798359580&password=12345678	1	0	100.00%	364 ms
对充电点发表一个评论 /api/pubPlugComment?pid=e82b3639c69fc786dce079382df14c01&content=api充电点评论&type=1	1	0	100.00%	330 ms
社区帖子列表 /timeline/list?catId=54fea61d217e3&time=0&type=all&limit=20	1	0	100.00%	162 ms
社区文字发帖 /api/pubPost?content=apiPost	1	0	100.00%	764 ms

▲图 9.36

9.5 JMeter 性能测试示例

本书仅介绍一个 JMeter 接口性能测试的简单示例。

JMeter 作为一个简单开源的基于 Java 的性能测试工具，使用起来也很简单。

JMeter 可以二次开发，复杂的情形可以自写代码，因此功能十分强大。

（1）建立测试计划—线程组。线程组下的线程数，也就是并发用户数，这里设置为 10，循环次数为每个并发用户的请求数，如图 9.37 所示。

▲图 9.37

（2）添加登录 HTTP 请求，添加登录接口相关的 IP、方法、路径、参数等，如图 9.38 所示。

▲图 9.38

（3）查看结果树，验证一下 HTTP 请求数据是否正常，如图 9.39 所示。

▲图 9.39

（4）添加聚合报告，如图 9.40 所示。

▲图 9.40

（5）单击运行按钮，本次测试结果及含义如图 9.41 所示。

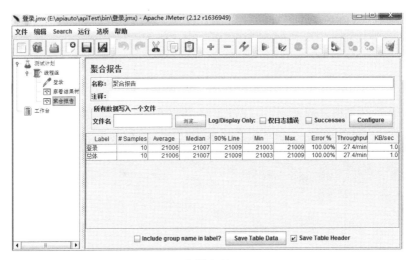

▲图 9.41

这里只有一个登录请求，因此在聚合报告中，显示一行数据，共 10 个字段，含义分别如下：

Label：每个 JMeter 的 Element（例如 HTTP Request）都有一个 Name 属性，这里显示的就是 Name 属性的值。

#Samples：表示这次测试中一共发出了多少个请求，如果模拟 10 个用户，每个用户迭代 1 次，那么这里就显示 10。

Average：平均响应时间——默认情况下是单个 Request 的平均响应时间，当使用了 Transaction Controller 时，也可以以 Transaction 为单位显示平均响应时间。

Median：中位数，也就是 50％用户的响应时间。

90% Line：90％用户的响应时间。

Min：最小响应时间。

Max：最大响应时间。

Error%：本次测试中出现错误的请求的数量/请求的总数。

Throughput：吞吐量——默认情况下表示每秒完成的请求数（Request per Second），当使用了 Transaction Controller 时，也可以表示类似 LoadRunner 的 Transaction per Second 数。

KB/Sec：每秒从服务器端接收到的数据量，相当于 LoadRunner 中的 Throughput/Sec。

第 10 章

LoadRunner 性能测试

10.1 小概念

LoadRunner 的小概念有很多，这里主要给出一些思路，不会一一详细解答，更多能是给大家一个知识学习的思考点。

1. 性能测试如何进行？

第一步：提取性能需求。例如，一个网站，有多少 VP、IP、并发用户、数据量、响应速度、吞吐率、带宽、环境配置？

例如，访问量趋势图，日均 PV80 万，独立 IP10 万，峰值 100 万；在线用户数、并发用户数、注册用户数：现在是多少？将来可能会是多少？系统所能支撑的情况……列一张表，然后与相关人员确认，商讨。

第二步：设计，计划，评审。

第三步：申请机器，环境搭建。

第四步：工具，脚本，造数据，场景执行，监控。脚本自己写还是录制？具体怎么做？如何增强优化？

第五步：结果分析，定位问题，性能优化，优化效果结果验证。

2. LoadRunner 将性能测试分为 6 个步骤：计划测试，测试设计，创建 VU 脚本，创建测试场景，运行场景，分析结果。

3. 在性能测试中，常见测试协议有 Web（http/html）、Socket、odbc、 WebService、SMTP。

4. LoadRunner 由哪些部件组成？各自用途是什么？

由 LoadRunner virtual user generator 、contorller running contorller 和 analysis 组成。

5. web_find 和 web_reg_find 的区别：后面的注册可以保存出现的参数次数。

6. 集合点的概念（集合点应插在事务之前）？什么是集合点？为什么要设置集合点？怎样设置？命令是什么？

集合点是用户数达到最大并发时同时操作，即向服务器发送请求数据而产生的压力，不设置可能起不到并发的效果。可加入集合点函数进行设置。

7. 一个 action 中有 3 个事务，如何让第一个事务运行 10 个虚拟用户，第二个事务运行 10 个虚拟用户，第三个事务运行 5 个虚拟用户。方法：设置集合点，集合点最大并发用户数分别设置为 10，10，5。

8. 响应时间连续升高，点击率、吞吐量下降，内存下降，CPU 一台占用 80%，另一台占用 20% 的原因是什么？

可能的原因：并发数量与响应时间成正比，并发越多，响应时间越长；页面假死会造成点击率下降；内存泄露会造成内存下降。

9. 响应时间升高到一个点后下降，点击率升高到最高后下降，这是为什么？

10. 参数化中取值的两种方式：

- 顺序，随机，唯一。
- 迭代取值，每次出现才取值，只取一次值。

11. 关联的函数和原理：web_reg_save_param()。

12. 压力测试结果分析：

- 网络瓶颈，局域网、外网带宽。
- 服务器硬件瓶颈。
- 服务器操作系统瓶颈。
- 中间件瓶颈（Web 服务器，数据库，参数配置）。
- 应用瓶颈（SQL 语句，数据库设置，业务逻辑，算法，代码等）。
- LoadRunner 脚本，参数，配置问题。

13. LoadRunner 测试出的结果中平均响应时间比真实时间长还是短？

14. LoadRunner 有浏览器功能么？

15. 一次测试结果发现，TPS（系统吞吐量）高得离谱，但是吞吐量和应用程序都没有任何消耗，问题出在哪里？

16．某次测试结果显示，并发用户、吞吐量、响应时间等曲线都很平稳，请问用这样的结果能分析出哪些性能问题？

17．如何实现让 LoadRunner 自动化连续跑 10 个不同的脚本？注意，不是一起跑，而是 10 个不同的场景，跑完后，怎样分析结果？

18．LoadRunner 一定要测试出问题么？性能测试的目的是什么？

19．性能测试主要有哪几种测试，各有什么目的？

20．性能测试与可靠性测试有何不同？

21．一个客户端有 100 个用户访问，与 100 个客户端有 100 个客户访问有什么区别？

22．性能优化有时就看几个指标：并发用户数、响应时间、事务成功率。如果这几个指标有一定提升，则不管开发人员修改什么代码，都算是性能优化。

23．性能测试 analysis 提供哪些报表？

24．场景设置有哪些方法：手动场景和面向目标的场景。

25．负载测试、容量测试和强度测试的区别？

负载测试：负载测试的主要目的是为了检测测试软件系统是否达到需求文档设计的目标。

例如，软件在一定时期内，最大支持多少并发用户数？软件请求出错率？响应时间？事务成功率等？测试主要是指测试软件系统的性能。

强度测试：强度测试就是压力测试。压力测试主要是为了测试硬件系统是否达到需求文档设计的性能目标。

例如，在一定时期内，系统 CPU 利用率、内存使用率、磁盘 I/O 吞吐率、网络吞吐量等。压力测试和负载测试的最大差别在于测试目的的不同。

容量测试：确定系统最大承受量。譬如系统最大用户数、最大存储量、最多处理的数据流量等。

26．响应时间和吞吐量是什么，有什么关联。

吞吐量显示的是虚拟用户每秒钟从服务器接收到的字节数。当和响应时间比较时，可以发现随着吞吐量的降低，响应时间也会降低。同样地，吞吐量的峰值和最大响应时间差不多同时出现。

27．什么是吞吐量？吞吐量是指单位时间内系统处理客户端的请求数。

28．PV 计算公式：

每秒*平均值 =((总 PV 量*80%)/(24*60*60*40%))/服务器数量(4) =pv/s= 8

每秒*峰值 = (((总 PV 量*80%)/(24*60*60*40%))*1.6) /服务器数量(4)=pv/s= 15

TPS=每台服务器每秒处理请求的数量=((80%*总 PV 量)/(24 小时*60 分*60 秒*40%)) / 服务器数量

29．Httpwatcht wireshark 工具的主要功能是：网页摘要、Cookies 管理、缓存管理、消息头发送/接受、字符查询、POST 数据和目录管理。报告输出能有效检测网页中各个元素的加载速度和加载时间，分析网页上隐蔽的下载链接，帮助发现网速被消耗在哪里。

30．yslow 前端页面性能

十分流行的雅虎网站优化规则给网站提出了十几条优化建议，包括尽可能地减少 HTTP 的请求数、使用 Gzip 压缩、将 CSS 样式放在页面的上方、将脚本移到底部、减少 DNS 查询等。

31．TPS（Transaction Per second，每秒钟系统能够处理事务或交易的数量）。

它是衡量系统处理能力的重要指标。

点击率（Hit Per Second）

点击率可以看作是 TPS 的一种特定情况。点击率主要体现用户端对服务器的压力,而 TPS 主要体现服务器对客户请求的处理能力。

每秒钟用户向 Web 服务器提交的 HTTP 请求数。这个指标是 Web 应用特有的一个指标;Web 应用是"请求—响应"模式,用户每发送一个申请,服务器就要处理一次,所以点击是 Web 应用能够处理的交易的最小单位。如果把每次点击都定义为一个交易,则点击率和 TPS 就是一个概念。可以看出,点击率越大,对服务器的压力也越大。点击率只是一个性能参考指标,重要的是分析点击时产生的影响。

需要注意的是,这里的点击并不等同于鼠标的一次"单击"操作,因为在一次"单击"操作中,客户端可能会向服务器发现多个 HTTP 请求。

32. pacing 值设置迭代之间的时间间隔:

pacing 值可用来调节 TPS,下面介绍它和 think time 的区别。

图 10.1 是 pacing 的三个选项,下面依次来说明。

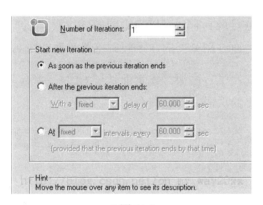

▲图 10.1

(1)第一项"As soon as the previous iteration ends",其含义是没有间隔,即 pacing 值不生效。

(2)如果选择第二项,如图 10.2 所示,其含义为: 当上次迭代结束后,等待一定时间后再进行下次迭代。

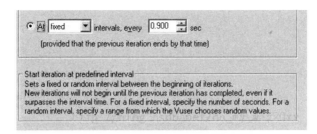

▲图 10.2

（3）如果选择第三项，则比较复杂，如图 10.2 所示的详述部分。

> 注意，如果选择这项设置：从上一次迭代开始到本次迭代开始的时间，则此时和响应时间的关系是：
>
> TPS = 1/pacing 值（这种模式的理想情况）
>
> 此种设置要保证 pacing 值大于 action 时间。

33．当不支持浏览器录制时，可以换不同的 IE 版本，或 Firefox、360、Opera 等浏览器，性能测试是通过浏览器的，不同浏览器对性能影响不大。

34．Web 前端性能测试：如果静态页面做过 CDN，那么怎么测试都没有意义，因为你访问的不是真实的页面，而是缓存。

要看缓存到哪里了：一个是缓存到本地，另一个是缓存到服务器。

35．Nmon 服务器资源监控。

软件安装包为 nmon_x86_centos6。

安装步骤如下：

（1）上传文件。

（2）权限 chmond 777。

（3）组件 yum install libncursesw.so.5。

（4）启动./nmon_x86_centos6。

（5）记录结果 ./nmon_x86_centos6 -fT -s 5 -c 10。

（6）打开文件 nmon analyser v334.xls。

36．在 VUGen 中何时关闭日志？何时选择标准和扩展日志？

在大型负载测试场景下，无须启用标准和扩展日志，而错误日志可以在调试和负载中使用。

37．性能测试准备：LoadRunner 客户机，带宽准备，服务器单接口（如 1000MB），交换器容量大小，有几个网口，是否绑定网卡，I/O 和网络传输。

开发设计处理能力多大，是否并发？响应时间？处理能力？成功率？业务模型有没有准备？用户场景有没有分析？

可能的性能瓶颈？系统？网络？中间件？数据库？应用？集群是哪种模式？软件是 LVS？硬件是 F5？

配置策略呢？LVS 有十几种策略，策略不一样，性能也大为不同。软件开发的架构？

Ngnix 有做优化么？JBoss 有做优化吗？GC 是怎样配置的？关闭防火墙试试？

如果是开发人员想做性能优化，有没有目标值？还是想看看是否有内存泄漏？导致崩溃的其他问题？或者设计上有没有心跳？守护进程?过载保护？数据冗余？一个集群 2 台是什么概念？是互为主备呢？还是并行运行？

38．socket 中的校验。

比如要校验和知道返回的第 30 个字符为用户名，第 35 个为账号，38 个为交易额；或者返回的字符是动态变化的。

校验时加参数，要用到关联。关联时，如果左边界或右边界为空是否可以？如果左右边界的值也是动态变化的，这时可以加入参数。

39．接口的性能测试。

Web_custom_request 方法可以发送 POST 和 GET 类型的请求。

Web_submit_data 只能发送 POST 类型的请求,所有 web_submit_data 方法发送的请求都可以使用 web_custom_request 来实现。

40．对产品的主要业务模块,如网站首页、会员登录、产品搜索与浏览、产品发布、产品询盘,实施性能测试并进行性能调优。

面试时问到性能方面,一般会问这个问题:我公司有 1 万人,有一个办公 OA 系统,现在需测试登录的压力,你是如何加压的?

(这个问题很不专业)你应该问他,我公司有 1 万人,每天使用 OA 10 万次,在 10 小时内,请你计算下每天的峰值峰谷是多少?

直接问并发多少的,这个属于什么级别?答案是还未入门的级别。

不问性能需求、业务模型、产品架构、业务流程、通信协议、开发语言和数据库、中间件,而直接告诉你并发 1000、500 直接开始测。

在测试开始前,一般先要分析出系统瓶颈可能出现的问题,在满足性能需求的前提下,需通过什么方案去开展测试,达成目标,从而发现预期的问题,而不是误打误撞,直接来个逐步加压法。不熟悉的人刚开始做也可以先尝试加压法,这是性能测试吗?当然不是,完全没有目标,撞上了问题只能算运气好。

性能测试分类有很多种,指标、压力、容量,还包括稳定性、可靠性,这是测试。还有性能调优,硬件、网络、系统、中间件、数据库的调优,都做了哪些?

性能是在一定环境下得出的定向结果,如果换了环境,则性能结果会有很大的不同。做性能测试,更要做好性能分析。

稳定性是在持续一段时间内(比如 72 小时)施加定量的压力,来评估系统的长效运行,比如成功率、时延、资源消耗、系统衰减等,可根据具体目标来施加相应的压力。

可靠性,目前可做的有很多,比如负载均衡、数据冗余、过载保护等。

针对过载保护的测试,就是要施加系统可承受意外的作用力来检测,在压力过大的情况下,系统是如何响应,如何处理业务的?当压力释放后,系统又是如何恢复运行的。在过载阶段导致的业务损失有多大？

41．获取性能指标。

根据性能需求文档、产品历史或相似版本的性能，分析客户数据、业界参考、性能测试人员的经验。用户关注的是用户操作的相应时间。

42．性能测试目的。

评估系统的性能，根据性能测试数据分析问题，优化性能指标，解决性能瓶颈和问题。

43．对 Web 做，优化脚本时，一般项是优化哪些？

缓存、思考时间、一大堆 URL 和图片链接等有些多余，可以删除；然后就是集合点、事务、参数化、关联、验证数据、日志等要加上。

44．测性能加压的时候，有没有同一用户同一 IP 同时多个操作的场景？如果没有，那么通常采用哪些场景？不同用户，不同 IP，同一时刻？

性能加压一般是在脚本里，设置不同的用户；而 IP，则在局域网设置一些 IP，虚拟一些 IP，这样才能模拟真实用户的场景。

45．逐个增加用户还是直接用最大值的用户数？

是逐步增加用户，这是在场景里设置的。首先脚本里设置好集合点，当场景运行时，在集合点就会产生并发，那每秒钟增加几个用户，持续运行一段时间，然后逐步释放用户。

46．假如有 100 个用户，那么我需要去做这么多用户还是复用一下？数据库里可能没有 100 个不同的用户。

最好做 100 个不同的用户名。

47．同一用户或者直接最大值用户数操作，这两种场景有实际意义吗？

一般情况下，没必要用同一用户或直接最大用户去做性能测试。

设置不同用户名的好处是，同一用户可能会产生错误，比如开发做了缓存。

缓存的意思是把这个用户名写到缓存中，再次访问时，直接从缓存里读取。

也就是说，客户端有 100 个同一用户并发，实际上在后台还是 1 个用户，因为用户名相同，所以缓存后被当作 1 个用户的请求处理了。

直接压最大值用户数，不是真实的用户场景。

我们是模拟用户真实的场景，直接最大值用户数，只是用户数加多一点，但不是绝对的。

比如考试系统并发，9 点开卷，1000 人考试可以设置用户数每秒增加 100 人，10 秒完成，这样就模拟了考生参加考试的场景，因为不太可能是 1000 个人同时点的。

48．逐步增加用户后要持续一段时间，这个时间有什么作用？以及怎样确定多少时间合适？

看稳定性，一般来说时间越长越真实，但是太长，会造成测试时间也很长，就不划算。合适的时间可以自己定，比如半小时、1 小时这样。如果是晚上，则可以设 8 个小时，第二天来看结果数据，工作时间有时半个小时就可以了。LoadRunner 默认是 5 分钟，有时也是可以的，有些脚本较为复杂，运行脚本的时间也比较长，这时最好持续时间长一些，不然会造成不协调或不准确。

10.2　安装

LoadRunner 11 试用安装包文件路径：https://pan.baidu.com/s/1qYxUqO8。

LoadRunner 是商业的，正式版需要向官方购买，因此本文所讲的都是供学习研究的试用版，如果在大企业请购买正式版。下载后单击 LoadRunner 11.0，运行"setup.exe"，如图 10.3 所示。

第 10 章　LoadRunner 性能测试　| 215

▲图 10.3

单击"安装"按钮，会提示缺少"Microsoft Visual C++ 2005 SP1 运行组件"下载这个组件，这里安装"vcredist_x86.exe"。安装完成后再一次运行"setup.exe"，安装程序会自动检查所需组件是否都已安装，确定都安装后会弹出欢迎界面，如图 10.4 所示。

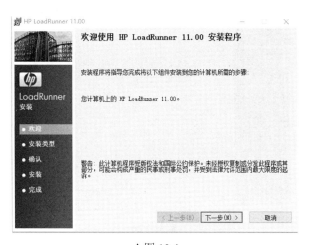

▲图 10.4

单击"下一步"按钮，勾选"Iaccept the termsin the tcense agreement"，如图 10.5 所示。

▲图 10.5

单击"下一步"按钮,注意安装路径不要包含中文。安装完成后的界面如图 10.6 所示。

▲图 10.6

单击"完成"按钮后,系统会自动打开"Loadrunner License Information"窗口,并提示你的 license 只有 10 天的使用有效期,单击 Close 按钮,关闭提示窗。

此时,就可以启动 LoadRunner 了。

10.3 脚本调试

LoadRunner 脚本录制。

start recording，选择 HTTP 协议，路径和浏览器可以默认，单击"开始"按钮，会弹出浏览器。打开默认要录制的网页，同时能看到有增加事件的提示 events，如图 10.8 所示。

▲图 10.8

录制时可在上面的工具栏。建 action，添加集合点，开始事务，结束事务，注释，检查点。

（1）登录流程选择脚本存放的 action。

vuser_int：首页脚本。

action：切换到登录界面的脚本。

增加 submit_login_action：提交登录的脚本。

vuser_end：退出登录的脚本。

（2）LoadRunner 脚本回放。

确认脚本的正确性，结合数据校验。

（3）LoadRunner 优化增强脚本。

- 通过工具栏或直接写代码的方式。
- 首先去掉缓存、思考时间，无用 URL、图片等脚本。
- 插入开始事务点 lr_start_transaction("start_login")。
- 插入结束事务点 lr_start_transaction("end_login")。
- 插入集合点 lr_rendezvous("rendezvous_login")。

- 插入检查点 web_reg_find("Text=成功登录",Search="body",LAST);。

优化脚本的重点和难点是参数化和动态关联。

1. 参数化

一般是静态数据，取值时需要取不同的值，既可以存储为 txt 文件，也可以在数据库里的

Parameter List 中新建一个参数，取参数为名 username，赋值 Mrzou。

在 Value＝处将 Mrzou 替换为 username，此时 Mrzou 会高亮显示为{username}。

为什么要参数化：如果数据库有缓存，如 redis、memcache 等，如果不参数化，即用同一个用户登录操作，则很可能会造成性能测试数据不准确，因为有缓存机制时，同一个用户名模拟 100 个用户，此时数据库的压力也只有一个用户，所以必须加上参数化，模拟不同的用户名密码进行登录操作，才能达到数据库服务器性能测试的真实测试效果。

2. 动态关联

关联（correlation）就是把脚本中某些写死的数据，转变成是取自服务器所送的、动态的、每次都不一样的数据，常见的需要关联的有 sessionId、token、cookies 等。

LoadRunner 中提供自动关联，也可以手动设置关联。脚本中掌握了参数化和动态关联，基本上工作中的绝大部分脚本都能关联，所以学好这个很有必要，其中关联还涉及正则表达式，也很简单，留意下左右匹配即可。

下面是关于参数化和关联的一个脚本示例。

3. 模拟 App 发送请求给云后台

模拟 App 发送请求给云后台

一般在用户登录后,云后台会返回登录成功的消息,并且返回一个 cookie 给 App。

当 App 下次要做一些例如设置名称之类的工作时，在请求消息里面会携带返回的 cookie，而且 cloud 也会校验这个 cookie。

（1）action 第一次请求登录，获取 cookie。

（2）action2 使用第一次请求获取的 cookie 进行第二次请求。

程序清单 10-1　c 代码

```c
Action()
{
//关联函数  动态参数变量为 par1，par2
web_reg_save_param("par1",
                    "LB=key\":\"",     //左匹配
                    "RB=\"",           //右匹配
                    LAST);
web_reg_save_param("par2",
 "LB=Set-Cookie: ",
 "RB=;",
 "ORD=2",
 LAST);

web_custom_request("login",
    "URL=http://192.168.10.10/api/user/login",
    "Method=POST",
    "Resource=0",
    "Referer=",
    "mode=HTTP",

    "Body=email={email}&password={password}",        //参数化 E-mail 和密码
                    LAST);

lr_message ("par1:%s", lr_eval_string("{par1}"));    //打印动态参数变量 par1 的值

lr_message ("par2:%s", lr_eval_string("{par2}"));    //打印动态参数变量 par2 的值

    return 0;
}

Action2()
{

web_cleanup_cookies();                               //头信息加入关联参数 par1，par2 的值
```

```
web_add_header("cookie",
    "{par1};{par2}");

web_custom_request("info",
    "URL=http://192.168.10.10/api/App/device",
    "Method=POST",
    "Resource=0",
    "Referer=",
    "mode=HTTP",

    "Body=caid=4026abc&key={par2}",                    //body 中加入 par2 关联参数的值
    LAST);
return 0;
}
```

如果是对接口（例如 App 的接口）进行性能测试，那么很可能没有脚本录制这一步骤，而是直接针对接口写 LoadRunner 脚本。

LoadRunner 通过 App 后台 Web 的 post 请求方法测试接口的性能

程序清单 10-2　c 代码

```
loginapi()
{
    web_url("rest",                                    //打开 HTML 的一个 URL 链接
        "URL=http://192.168.117.154/router/rest",
        "Resource=0",
        "RecContentType=text/html",
        "Referer=",
        "Snapshot=t1.inf",
        "Mode=HTML",
        LAST);
    web_url("favicon.ico",                             //打开一个 favicon 的一个 URL 链接
        "URL=http://192.168.117.154/favicon.ico",
        "Resource=0",
        "RecContentType=text/html",
        "Referer=",
        "Snapshot=t2.inf",
        "Mode=HTML",
        LAST);
```

```
    lr_rendezvous("loginapi");                          //增加的集合点,注意:通常集合点要放在事务的前面
    lr_start_transaction("loginapi");                   //增加开始事务
    lr_set_debug_message(LR_MSG_CLASS_EXTENDED_LOG | LR_MSG_CLASS_RESULT_DATA, LR_SWITCH_ON );
//调试日志信息开关
    web_reg_find("Text=操作成功",                        //设置检查点
        LAST);
    web_submit_data("testpost",                         //发送 POST 接口的请求数据
                "Action=http://192.168.117.154/router/rest",
                "Method=POST",
                "RecContentType=text/html",
                "Mode=HTML",
                ITEMDATA,
    "Name=method","Value=test.user.UserService.userInfoByTicket",ENDITEM,   //第一个参数和值
    "Name=v","Value=1.0",ENDITEM,                       //第二个参数和值
    "Name=App_key","Value=120",ENDITEM,                 //第三个参数和值
    "Name=ticket","Value=11111111111111111111111111111111",ENDITEM    //第四个参数和值
    "Name=sign","Value=22222222222222222222222222222222",ENDITEM,     //第五个参数和值
    LAST);
    lr_set_debug_message(LR_MSG_CLASS_EXTENDED_LOG | LR_MSG_CLASS_RESULT_DATA, LR_SWITCH_OFF);
//调试日志信息开关
    lr_end_transaction("loginapi", LR_AUTO);            //增加结束事务
    return 0;
}
```

LoadRunner 验证脚本,单击回放按钮,全部绿色打勾表示基本是成功的,如果有错误,就根据错误信息分析和解决。

10.4 运行场景

LoadRunner 运行。

场景设置:单一场景,复杂场景,基线性能。

单一场景设置:比如登录接口压服务器、即时消息接口压服务器等。

复杂场景:根据用户的操作习惯进行模拟,比如有 100 个用户,其中 30 个用户登录,50 个用户查看某单个商品,20 个用户下订单,这时我们既可以写一个脚本执

行，也可以写三个脚本执行。

基线性能：比如线上运行环境，用一两个用户时的性能指标，作为性能测试指标的参考基准，进行比较后再去优化。

场景运行：通常情况下，先每秒钟增加几个用户，到最大并发用户集合点后，再持续运行一段时间，比如默认为 5 分钟，或持续运行一个晚上也可以，然后每秒钟退出几个用户，直到场景运行结束。场景运行完成后，通过监控以及 LoadRunner 生成的测试报告数据进行记录、分析、定位、调优。

大量用户并发：当同时执行并发用户操作运行场景时，一台 LoadRunner 客户端机器的硬盘、网络、CPU、内存等资源是有限的。例如，我们要做 2000 个虚拟用户并发，可能 1 台 LoadRunner 运行 20 个用户并发时，CPU 或内存就耗尽了，因此需要装 100 台 LoadRunner 客户端（装客户端可以用操作系统克隆的方法，以节省时间和资源），每台设置 20 个用户并发，支持这 2000 个虚拟用户并发。这时可以设置一台主控机 LoadRunner，同时控制连接另外 99 台装有并启动 LoadRunner 的 agent 客户端，来执行这个场景，以保证测试的有效性。客户端节点加载图如图 10.9 所示。

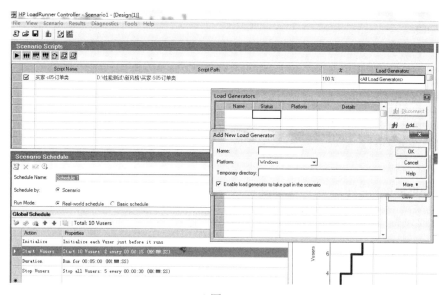

▲图10.9

10.5 性能监控

性能测试通常需要监控的指标如下。

- LoadRunner 性能工具：并发用户数，事务响应时间、TPS 等。
- LoadRunner 机器资源：CPU、内存、网络、磁盘空间等。
- 服务器 Linux 机器资源：CPU、内存，磁盘 I/O 等。
- 中间件：1.JBoss 2. Apache 3.Ngnix 等。
- 数据库：缓存命中、索引、SQL 性能、数据库线程数、连接池等。
- 网络：带宽、吞吐量、吞吐率等。

性能测试通常采用的监控方法如下。

1．Nmon 工具

全面监控 Linux 系统资源的使用情况，包括 CPU、内存、I/O 等，可独立于应用监控。

Nmon 的搭建过程如下：

（1）上传文件

（2）权限 chmond 777

（3）组件 yum install libncursesw.so.5

（4）启动 ./nmon_x86_centos6

（5）记录结果 ./nmon_x86_centos6 -fT -s 5 -c 10

（6）打开文件 nmon analyser v334.xls

2．Spotlight 工具

能通过图形直观动态显示硬件资源监控，并有警报信息等，界面如图：

3．常见命令

常见命令有 top、vmstat 等。

10.6 问题分析和调优

1．性能测试分析

从代码、服务器、数据库、网络等进行性能瓶颈分析。

2．性能分析数据

性能分析数据包含（1）LoadRunner 测试数据 TPS、响应时间、用户并发数等；（2）运维监控到的服务器、网络、数据库等数据；（3）应用服务器错误日志、超时日志等；（4）DBA 提供的数据库监控数据，开发人员提供的系统架构或代码逻辑等。

3．性能标准参考

TPS 是否大于期望高峰值？

CPU 利用率是否小于 75%？

内存使用率是否小于 80%？

错误概率是否小于 0.01%？

页面 YSlow 评级是否是 C 级以上？

页面响应时间是否<3 秒或 5 秒？

4．性能常见问题

性能常见问题有代码内存泄漏、SQL 语句、索引、中间件、并发量、防火墙、网络带宽、硬件配置等。

并发用户多时会导致数据库死锁。此外，性能常见问题还包括数据库配置问题，Nginx、JBoss 问题等。

另外，前端响应时间长、页面内容过大、图片过大、图片没有采用异步加载等也属于性能常见问题。

5．性能场景示例

例如，某系统派发红包，准备发 100 万元红包给用户，结果发了 150 万元的红包。其原因是，当大用户量同时抢红包时，比如 1 个用户抢走 1 个红包，数据库应该减少 1；2 个用户同时抢走一个红包，数据库应减少 2，但是由于大用户量同时抢红包时出现性能问题，导致数据库只减少了 1，所以统计的数字比抢到的数字少，从而多发出了红包，造成损失。

很典型的例子还有 12306 最早期上线时大家所熟知的性能问题。

6．性能调优

开发和运维采用缓存、降级策略、限流排队、增加服务器等比较基本的方法。

性能测试、代码调优、数据库调优、服务器调优等是比较有价值且经济的方法，这也是性能自动化测试人员的价值所在。

10.7　性能压力测试报告样例

1．概述

为满足系统现场大数据量、多用户并发登录及各种应用场景，使服务器和系统能稳定正常运行，因而需要做压力测试，最终根据测试结果数据反映系统的支撑能力。

该报告详细给出服务器优化之后，相关接口的具体性能。

2．需求

本系统在一定的测试环境软硬件配置及网络下，支持如下参数：

- 并发登录：100 用户并发登录。
- 即时通信：100 用户并发通信。
- 业务接口：100 用户并发获取帖子列表、预约试驾、资讯列表、文件上传。

3．目标

（1）构建与实际环境相匹配的基础数据环境。

（2）系统性能需求设计性能测试方案，定义业务模型及测试场景。

（3）对比各参测系统的性能指标，制作综合评测报告，为评测系统性能及性能优化提供参考依据，检验系统上线前，程序是否有大的并发漏洞和性能瓶颈。

4．测试环境

服务器网段与 IP 分配表、硬件。

5．角色和职责

部门和角色	职责	人员

6．测试场景方案

（1）基线测试场景。

基线测试场景用来验证系统功能完整性和可用性,以及测试脚本的可重复性。

序号	功能名称	功能点	并发	循环次数	间隔
	登陆	CAS登录鉴权	1	1	0
		API登录	1	1	0
		即时通讯登录	1	1	0
	即时通讯	即时通讯发消息	1	1	0
	业务接口	获取帖子列表	1	1	0
		预约试驾	1	1	0
		资讯列表	1	1	0
		文件上传	1	1	0

(2)单接口场景。

单接口测试场景主要是为了检验各接口模块是否有严重的性能障碍,检验性能处理能力的最大值。

序号	功能名称	子序号	并发人数				循环时间
	CAS登录鉴权	1	10	30	50	100	5分钟
	API登录	2	10	30	50	100	5分钟
	即时通讯登录	1	10	30	50	100	5分钟
	获取帖子列表	1	10	30	50	100	5分钟
	预约试驾	2	10	30	50	100	5分钟
	资讯列表	3	10	30	50	100	5分钟
	文件上传	4	10	30	50	100	5分钟
	合计						

用每秒钟增加5个用户,场景结束时每秒钟退出5个用户,持续循环运行5分钟

(3)混合接口场景。

按照分析的业务模型,对不同操作按照一定比例,分别执行不同压力下的测试场景。

(4)测试场景。

采用100用户并发,持续执行8小时。

(5)测试结果分析。

(6)问题和建议。

(7)附件。

性能测试详细数据记录1-4

1. 即时通讯服务器并发登录时的性能测试数据以及趋势图

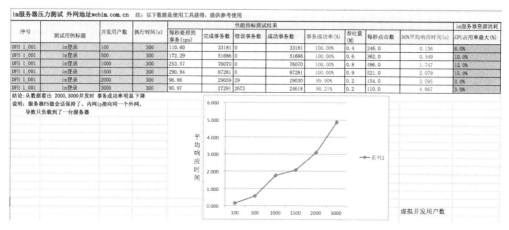

▲图 10.7

2. 鉴权服务器并发登录时的性能测试数据以及趋势图

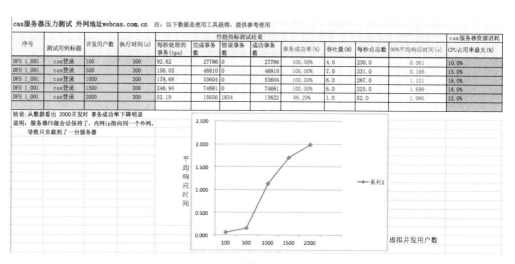

▲图 10.8

3. 接口服务器并发登录时的性能测试数据以及趋势图

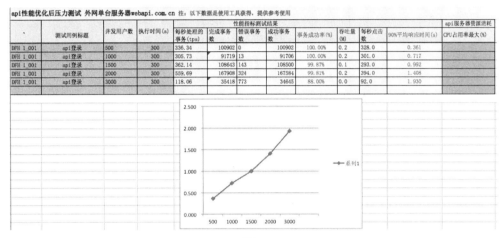

▲图 10.9

4. 文件服务器并发上传文件时的性能测试数据以及趋势图

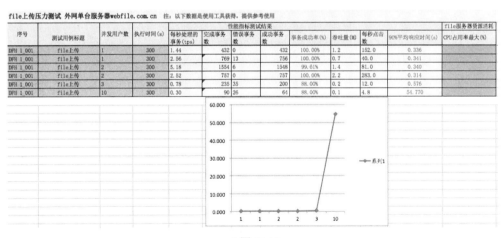

▲图 10.10

性能压力测试结果趋势图 1-3

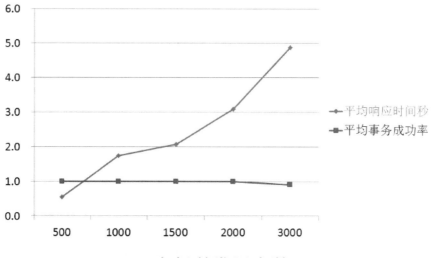

1. IM虚拟并发用户数

▲图 10.11

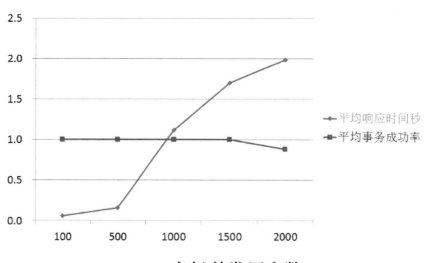

2. CAS虚拟并发用户数

▲图 10.12

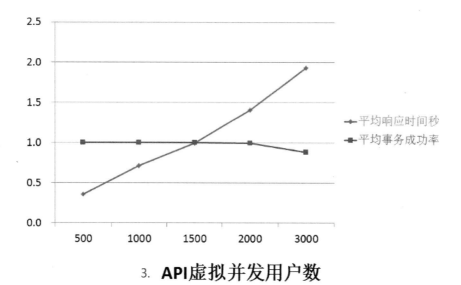

3. **API虚拟并发用户数**

▲图 10.13

第 11 章

Jenkins 持续集成

11.1 介绍

Android 自动化、iOS 自动化、API 自动化、Selenium 自动化，JMeter 和 LoadRunner 性能测试脚本，这些全都可以集成到 Jenkins 平台，加上本书第 1 章介绍的用 Ant 和 Svn 进行自动化部署、编译、运行、测试结果自动邮件通知，从而实现无人值守全程自动化，前面章节已有介绍，这里不再赘述。

Tomcat 启动 startup.bat 后，把 jenkins.war 包放到 webApp 目录下，Jenkins 即可运行。或用 Jenkins 1.6 安装包，直接安装后即可运行。

安装文件路径：https://pan.baidu.com/s/1b0OzXs。

11.2 系统配置

1. 系统管理（Configure Global Security），设置全局变量，勾选允许用户注册，如图 11.1 所示。

▲图 11.1

2. 注册账号并登录。

▲图 11.2

3. 设置 JDK。

▲图 11.3

4. 设置 Ant。

▲图 11.4

5. 设置邮件。

▲图 11.5

6. 管理插件。

可选插件中先安装如下要用到的插件。

- 邮件插件：Regex Email Plugin。
- TestNG 插件：TestNG Results Plugin。
- Ant 插件：Ant Plugin。
- SVN 插件：Subversion Plug-in。
- 报表插件：ExtentReports。

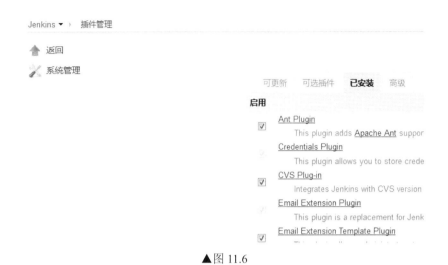

▲图 11.6

7. 管理节点，新建本地节点并连接。

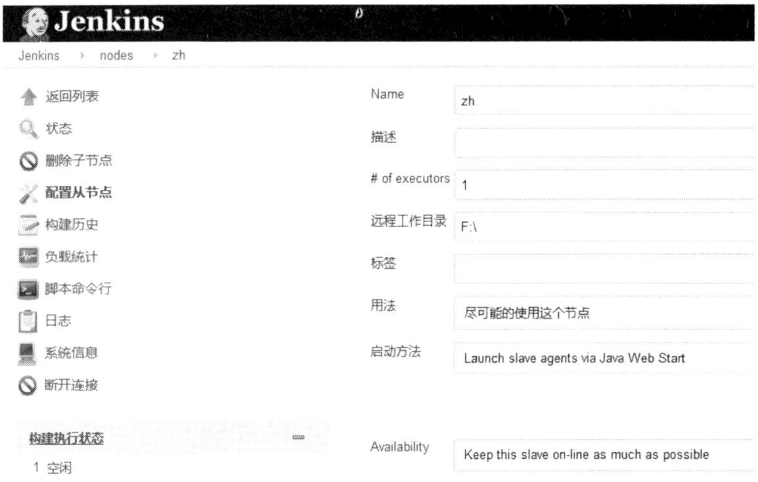

▲图 11.7

11.3 项目配置

增加 project 构建项目，包括上述的 Android 与 iOS 自动化、API 自动化、Web 自动化、JMeter 自动化、LoadRunner 性能自动化项目。

1. 在视图 App_autotest 中新建 job 项目。

▲图 11.8

2. 所运行的客户端机器为 zh。

▲图 11.9

3. SVN 设置如图 11.10 所示。

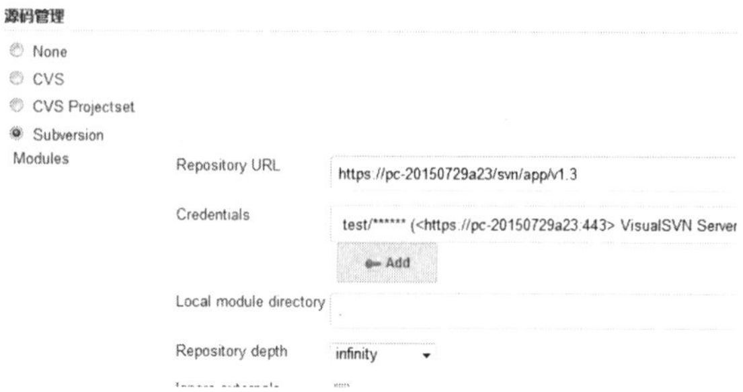

▲图 11.10

密码验证如图 11.11 和图 11.12 所示。

▲图 11.11

▲图 11.12

4．触发器设置，这里选 Build Periodically（自动周期执行），时间为周一到周五早上 10 点。也可以设置第 2 个 Build after other projects are built，即在某一项目执行后再触发此项目，如图 11.13 所示。

▲图 11.13

5. TestNG report 在 Jenkins 上输出报告。

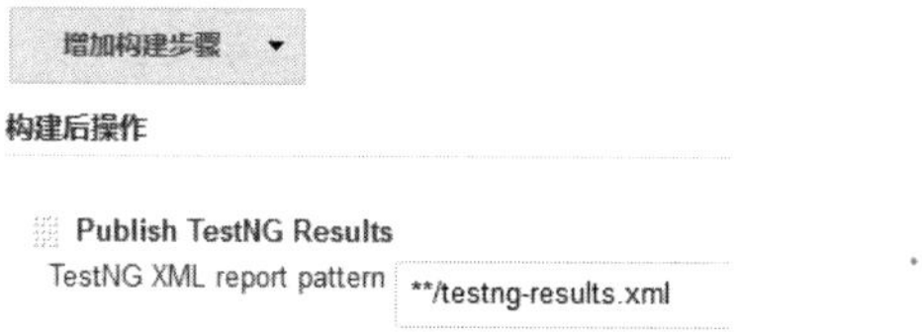

▲图 11.14

6. E-mail 自动邮件配置，邮件发送的内容设置如图 11.15 所示。

▲图 11.15

7. API 自动化项目中的 Python 脚本如图 11.16 所示。

构建

Execute Windows batch command

命令
```
python testAPIbuyerV1_2.py
python OrderSubmitFlowV1_2.py
python OrderCancelFlowV1_2.py
python OrderDelFlowV1_2.py
python OrderPayFlowV1_2.py
python OrderRefundFlowV1_2.py
python OrderDeliverFlowV1_2.py
python OrderCancelRefundFlowV1_2.py
python OrderReturnDeliverFlowV1_2.py
python OrderCancelReturnDeliverFlowV1_2.py
python testAPIbizV1_2.py
python testAPIbizPhysicalV1_2.py
python bizOrderDetailV1_2.py
ping -n 10 127.0.0.1>nul
python bizOrderDeliveryV1_2.py
```

▲图 11.16

11.4 多机器节点配置

1. 节点 slave 配置及连接，可以设置多台机器节点：

http://192.168.115.146:8080/jenkins/computer/

增加子节点，单击系统管理→管理节点→新建节点，输入节点名称，单击 OK 按钮，保存该节点。配置节点如图 11.17 所示。

第 11 章 Jenkins 持续集成

▲图 11.17

远程工作目录，即节点电脑从 SVN 传自动化脚本的目录，启动方法一般如图 11.18 所示。

▲图 11.18

2. 连接子节点，单击"launch"按钮，或新建.bat 文件。

▲图 11.19

如果出现启动程序失败,则需在 Firefox 下直接单击"Launch"按钮,选择 Java 运行。

连接后不要单击"install service",否则会在后台运行测试用例。

▲图 11.20

3. Job 和节点机器关联:Restrict where this project can be run。

▲图 11.21

4．启动远程节点自动化，为多台机器配置远程节点 JDK、Ant 环境、XP 系统，如果构建出现 test flied，则需要查看启动的浏览器安装及路径是否和程序中的一致。

11.5　结果展示视图

1．Jenkins 展示项目视图（各个自动化测试项目）。

▲图 11.22

2．TestNG 展示测试报告 （每次构建的结果趋势图）。

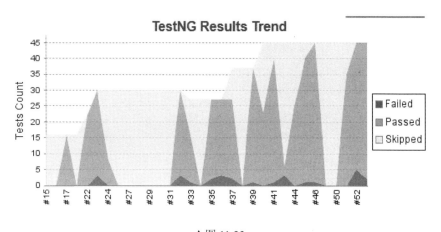

▲图 11.23

3. App 自动化测试用例执行构建日志（下面是 45 个用例执行通过）。

```
[testng] 2016-08-24 15:08:08,569 INFO main [com.netease.qa.testng.TestResultListener]onFinish(line:56)PassedTests = ZiYing_Logout_001
[testng] 2016-08-24 15:08:08,569 INFO main [com.netease.qa.testng.TestResultListener]onFinish(line:56)PassedTests = Home_Store_Weixin_001
[testng] 2016-08-24 15:08:08,569 INFO main [com.netease.qa.testng.TestResultListener]onFinish(line:56)PassedTests = FenXiao_Login_001
[testng] 2016-08-24 15:08:08,569 INFO main [com.netease.qa.testng.TestResultListener]onFinish(line:56)PassedTests = Home_Incoming_Tixian_001
[testng] 2016-08-24 15:08:08,570 INFO main [com.netease.qa.testng.TestResultListener]onFinish(line:56)PassedTests = Upgoods_Recommend_001
[testng] 2016-08-24 15:08:08,570 INFO main [com.netease.qa.testng.TestResultListener]onFinish(line:56)PassedTests = Home_DelGoods_Classify_001
[testng] 2016-08-24 15:08:08,570 INFO main [com.netease.qa.testng.TestResultListener]onFinish(line:56)PassedTests = Home_Store_Qualification_001
[testng] 2016-08-24 15:08:08,570 INFO main [com.netease.qa.testng.TestResultListener]onFinish(line:56)PassedTests = FenXiao_Logout_001
[testng] 2016-08-24 15:08:08,570 INFO main [com.netease.qa.testng.TestResultListener]onFinish(line:56)PassedTests = Home_DownGoods_Classify_001
[testng] 2016-08-24 15:08:08,570 INFO main [com.netease.qa.testng.TestResultListener]onFinish(line:56)PassedTests = Home_Store_Renovation_001
[testng]
[testng] ===============================================
[testng] Suite
[testng] Total tests run: 45, Failures: 0, Skips: 0
[testng] ===============================================
[testng]
transform:
    [xslt] Processing F:\workspace\star-app-v1.3\test-output\testng-results.xml to F:\workspace\star-app-v1.3\test-output\Report.html
    [xslt] Loading stylesheet F:\workspace\star-app-v1.3\lib\testng-results.xsl
BUILD SUCCESSFUL
Total time: 27 minutes 36 seconds
TestNG Reports Processing: START
Looking for TestNG results report in workspace using pattern: **/testng-results.xml
Saving reports...
Processing 'F:\Program Files (x86)\Jenkins\jobs\star-app-v1.3\builds\59\testng\testng-results.xml'
TestNG Reports Processing: FINISH
Email was triggered for: Always
Sending email for trigger: Always
Sending email to:
Finished: SUCCESS
```

▲图 11.24

附录 A

自动化管理平台和产品自动化系统

1. 企业级自动化测试管理平台架构图

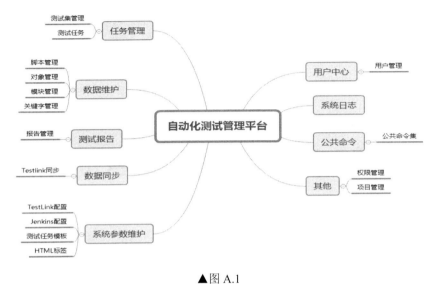

▲图 A.1

2. 企业级产品自动化测试系统功能图

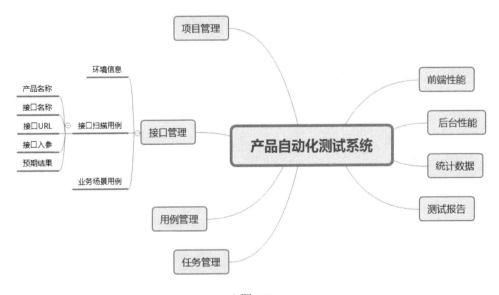

▲图 A.2

附录 B

Java 和 Python 开发语言学习历程

1. Java 语言的学习历程

记得最初找第一家工作的时候,是面试开发,面试官看了我的答题后,第一句话就对我说,让你做测试你做不做?我毫不犹豫地说,做。

接着她问我冒泡算法、语言、数据结构等一些大学里面的编程语言题目,基本上我也都回答出来了,实际上我的语言基础很是一般,是由于面试前做了些题目,就像大学时临考前的看重点一样,所以才回答出这些题目,因此做一些企业需要的技能准备是很有必要的。

于是我就进了这家企业。这是一家外包企业,给华为和平安做项目,面试我的是个女博士,管理很严厉。上级是个跟我体格相似的,因此还算比较照顾我。

当初的项目非常的规范,开发的源代码我们也是可以看的,我没事儿时就看这个用 Java 做的项目,始终看不懂,实际上并不要求测试人员能看懂代码。做了一段时间测试后,我还是有转开发的想法。

详细就不介绍了,简单地讲述我的 Java 语言学习历程,大概就是最初工作一两年中,一直想转开发,然后下班宅在家模拟源代码和书籍,完整的一行一行地敲代码,写了简单购物网站的前后台,大概历时 6 个月。

在第三年的时候接触参与了华为的报价配置器项目的自动化工具开发,原本是 RFT 进行自动化,华为要把它转换成 Java 版的自动化工具,由于我有代码基础,因此有机会参与了该项目,大概历时 3 个月的时间。

在走秀网,由于大用户量增长,从而导致电商网站响应时间特别慢长,此时正式接触和研究学习并实践性能测试,大概历时 9 个月。

到华南城时由于当上了主管职位,于是一些功能测试任务就给相关的初、中级测试人员,这时我重点研究自动化测试开发技术。

最近两年工作中是全程专职负责自动化,主要是接口自动化和 App 功能自动化,现在可以说是得心应手了。

从我个人经验以及相关性能和自动化测试开发从业人员的经验可以知道,掌握一门开发语言是基础，是必不可少的前提条件。

　　Java 是一门面向对象语言（具体的学习可以查找相关开发语言官方资料，实践是王道）。

　　第一节 语法，8 种数据类型。

　　第二节 类、对象、方法、变量、常量。

　　第三节 继承、封装、多态、重写、数组、集合、包，接口、反射、文件等。

小实例练习

　　（1）hello world 代码小实例（hello world 的成功运行是学习的第一小步）

```java
public class HelloWorld //一个文件中只能有一个共有的类，文件名称大小写必须一致
{
    public static void main(String[] args) //程序入口
    {
        System.out.println("HelloWorld！");//系统函数向控制台输出 Hellowrold
    }
}
```

　　（2）myfun 加减运算小实例

```java
public class MyFun
{
public int sum(int a,int b)    //加法运算的方法
{
    int resultSum=a+b;         //定义变量并实现 2 个参数值相加
    return resultSum;          //返回变量值
}

public static void main(String[] args)
{
    MyFun test= new MyFun();     //创建类的对象并初始化
    System.out.println(test.sum(1,1));   //输出对象方法的参数加法运算
}
}
```

2. Python 语言的学习历程

我是在工作中学到的 Python 语言，当时除了 8 个功能测试同事外，分别还有 1 个后端硬件接口自动化测试，1 个后端 App 接口自动化测试和 1 个性能测试，而我负责 App 自动化测试。分工比较明确，但是大家的开发语言各不相同，一个用 Python，一个用 perl，一个用 C，我用 Java。后来经过沟通，大家互相培训讲解各自的语言和自动化测试思路并演示，再后来了解到 Python 是最流行的脚本语言，因此同事们纷纷都选择 Python 语言进行学习以及尝试接口测试，我也不例外。其中一个同事来自百度，也提到他之前公司全是自动化测试开发人员，当开发人员不够时可以从测试开发人员转岗，这在一定程度上启发了我的构书思路，足以说明自动化测试开发技术在企业中是一个趋势，是有一定价值的，而我正好掌握和接触过这些技术，于是把整套自动化测试开发相关的东西写出来，非常有意义，不是么？

Python 是解释型脚本语言（具体的学习可以查找相关开发语言官方资料，实践是王道）。

第一节 欢迎来到 Python 世界，什么是 Python？

第二节   Python 基础。

第三节   数据类型。

第四节   序列。

第五节   映像类型。

第六节   流程控制。

第七节   函数等。

小实例练习

（1）hello world 代码实例（hello world 的成功运行是学习的第一小步）

```
#encoding: utf-8
"Filename : helloworld.py"      //python 保存文件命名
def helloworld():               //定义函数
   print "HelloWorld"           //输入 Hello World
```

```
if __name__=='__main__':#__name__是指示当前py文件调用方式的方法
    helloworld()
```

（1）myfun 加减运算小实例

```
#encoding: utf-8
"Filename : myfun.py"
class test(object):
    def myfun(self,x,y):        //定义加法函数
        print x+y               //输出参数变量的加法运算值
    def myfun2(self):
        pass
if __name__=='__main__':#__name__是指示当前py文件调用方式的方法
a=test()                    //定义变量
a.myfun(1,1)                //调用加法运算函数
```

附录 C

常见错误和问题解答

由于在实践过程中，会碰到各种各样的问题，因此本节的目的在于总结出一些常见问题以及解决办法，大家在实践过程中可以自己进行积累、总结、记录，以下仅供参考。

1．版本不对应引起的各种问题

解答：将 Windows 操作系统 32、64 位，JDK 版本，Eclipse 版本，LoadRunner 版本，浏览器版本，SDK 版本，ADT 版本，iOS 版本，Xcode 版本，Appium 等改成一致的正确的版本即可。

2．Appium 不能输入中文

解答：请看 Android_config.properties 中的代码解析。

3．运行脚本时直接报错

解答：adb devices 手机连接，手机调试模式打开，启动 Appium server。

4．控件元素找不到

解答：有些元素 id 或 name 重复，或者 id 和 name 为空，可以要求开发人员添加；或者就是抓取控件元素时弄错了，还有可能是页面元素加载还没完等等。

5．接口测试脚本运行时数据库连不上

解答：root 远程权限，接口测试脚本中 localhost 或 127.0.0.1。一定要换成电脑的 IP 地址。

6．LoadRunner 遇到的问题，比如 IE 打不开、录制不到、安装、破解失败等

注意安装过程中的操作系统和浏览器，很多时候都不太兼容，这时，要查找相关资料，根据错误提示判断并解决它。LoadRunner 11 兼容 Win 7、Win 2008、IE 8、IE 9 等。另外，注意电脑系统是 32 位还是 64 位，都要对应检查。

LoadRunner 录制脚本经常遇到不能打开浏览器的情况，可以用下面的方法来解决。

启动浏览器，打开 Internet 选项对话框，切换到高级标签，去掉"启用第三方浏览器扩展（需要重启动）"的勾选，然后再次运行 VuGen 即可。

提示：通常安装 Firefox 等浏览器后，都会勾选上面的选项，导致不能正常录制。运行 LoadRunner 的主机上保持一个干净的测试环境，用管理员权限安装和运行，关闭和删除杀毒软件，关闭防火墙之类等不兼容性软件。

附录 D

常用软件安装包链接

1. 官方网站汇总链接地址

Eclipse：http://www.eclipse.org/downloads/

JDK：http://www.oracle.com/technetwork/Java/Javase/downloads/index.html

SDK：http://Android-SDK.en.softonic.com/

ADT：http://tools.Android-studio.org/index.php

Node.js：https://nodejs.org/en/

TestNG：http://TestNG.org/doc/index.html

Ant：http://ant.apache.org/

Appium：http://Appium.io/

Jenkins：https://jenkins.io/index.html

Python：https://www.Python.org/downloads/

Pydev：http://www.pydev.org/download.html

Postman：https://www.getpostman.com/

SVN：https://tortoisesvn.net/downloads.html

Zentao：http://www.zentao.net/

MySQL：https://www.mysql.com/

2. 百度云盘汇总链接地址

（1）Android

JDK 1.7 安装文件路径：　　　　　　https://pan.baidu.com/s/1gf4Ym3L

Eclipse 4.5.2 安装文件路径：　　　　https://pan.baidu.com/s/1dF0sBcP

TestNG 安装文件路径：　　　　　　https://pan.baidu.com/s/1bLhluA

SDK r24.4.1 安装文件路径： https://pan.baidu.com/s/1mi6PT9m

ADT 安装文件路径 23.0.7： https://pan.baidu.com/s/1sl2BZit

Node.js x64 安装文件路径： https://pan.baidu.com/s/1pKLwEFp

.Net 4.5 安装文件路径： https://pan.baidu.com/s/1sl1qdgL

Appium.exe v1.4.13 安装文件路径： https://pan.baidu.com/s/1jHGhnxG

（2）iOS

Appium.dmg v1.4.13 安装文件路径： https://pan.baidu.com/s/1jIJfruA

JDK 1.8 macosx x64 安装文件路径： https://pan.baidu.com/s/1i5nZl0D

（3）API

Postman osx 4.5.1 安装文件路径： https://pan.baidu.com/s/1jIRSI70

Python 2.7.10 安装文件路径： https://pan.baidu.com/s/1skBPj0h

Fiddler v2.2.2.0 安装文件路径 https://pan.baidu.com/s/1o7USimA

Postman 4.1.2 安装文件路径： https://pan.baidu.com/s/1jICtHP0

（4）Zentao 安装文件路径

Navicat for MySQL 安装文件路径： https://pan.baidu.com/s/1slb8boh

PyDev 4.5.4 安装文件路径： https://pan.baidu.com/s/1jH9vBmu

MySql-python-1.2.3 安装文件路径： https://pan.baidu.com/s/1o8B58p0

Requests-2.11.1 安装文件路径： https://pan.baidu.com/s/1dF1E3Wt

（5）Web

selenium-server-standalone-2.39 文件路径 https://pan.baidu.com/s/1o8Cl6oq

selenium-java-2.43.1-srcs 安装文件路径 https://pan.baidu.com/s/1qXA5s6K

selenium-java-2.43.1 安装文件路径　　　　https://pan.baidu.com/s/1slJ6VJB

IEDriverServer 安装文件路径　　　　　　　https://pan.baidu.com/s/1i5BFMJV

chromedriver 安装文件路径　　　　　　　　https://pan.baidu.com/s/1miO0QG4

testng.jar 安装文件路径　　　　　　　　　　https://pan.baidu.com/s/1hsbjZt6

saxon-8.7.jar 安装文件路径　　　　　　　　https://pan.baidu.com/s/1gf2qwNL

Jmeter：

Jmeter 安装文件路径：　　　　　　　　　　https://pan.baidu.com/s/1kVJdnuv

Loadrunner：

LoadRunner 11 安装包和破解文件路径　　　https://pan.baidu.com/s/1qYxUqO8

LoadRunner 12.0 安装文件路径　　　　　　https://pan.baidu.com/s/1miLddiw

Nmon 安装文件路径：　　　　　　　　　　 https://pan.baidu.com/s/1eSa9sbg

bwm-ng 安装文件路径：　　　　　　　　　 https://pan.baidu.com/s/1boS6G7X

Spotlight for MySQL 安装文件路径：　　　　https://pan.baidu.com/s/1jItwN5s

Spotlight for UNIX 安装文件路径：　　　　　https://pan.baidu.com/s/1nvlPket

（6）Jenkins

Ant 1.9.7 安装文件路径：　　　　　　　　　https://pan.baidu.com/s/1c1IvthY

SVN 安装文件路径：　　　　　　　　　　　https://pan.baidu.com/s/1nvB7BhZ

Jenkins 1.64 安装文件路径：　　　　　　　　https://pan.baidu.com/s/1b0OzXs

App 自动化测试框架 demo 源码包：https://pan.baidu.com/s/1bEwMh4　　密码：g7pk

API 自动化框架 demo 包：https://pan.baidu.com/s/1eSDiVhK　　　密码：bjbx

Web 自动化框架 demo 源代码包：https://pan.baidu.com/s/1kV4Rkvd　　密码：4rfu

后　记

祝我们每个人都能够收获属于自己的价值，不仅包括技术、职位、金钱，还有生活和其他，做越来越强大的自己！

我们需要不断提升自身价值，也就是所说的身价。健康、好奇心、有梦想、坚强、自律和自省，主动学习新的知识，对不确定性保持乐观，不甘平庸，对重要事情有判断力，有很好的计划性和执行力等优秀的品质。从小我们就被教育成听父母的话，听长辈的话，听老师的话，自然而然潜移默化地形成了要他人立规矩，形成了奴性。常常努力争取他人的认可和同意，太在意他人的看法，循规蹈矩，做个老实人，做个好人，为我们的邻居、男神女神、同事、上级，等等着想。但是他们一旦不予好的回馈，我们就觉得自己变得一无所有、消极忧郁，或者给以反对，我们更是心灰意冷，不能坚持己见，丧失自我，顺从，依赖，最后责怪抱怨他人。当我们没有价值时，比如乞丐流浪汉又如何爱别人呢。因此要先自爱，做自己的主人，先学会照顾好自己的提前下再爱他人。当我们自爱，自身才会成长、健康、独立、体贴、坚强，既便没有别人的称赞和掌声，也一样积极的、喜悦的、自爱的工作生活，在生活中、在工作中不刻意求别人关注，也不处处谦让别人、自卑软弱受欺负、而自己不求进取。

软件自动化测试开发技术，目前算属于测试行业比较制高点，随着大家技术提升的同时，建议一定要将其应用到项目业务实践中发挥价值，否则技术再牛也是个 0，毫无意义。我们功能测试，依我个人从业经历来看，三个月的积累时间足以，撑死半年，其他时间都是重复。甚至有些转测试的人压根就没有测试知识，依然在从事功能测试工作。专业一点讲功能测试虽然重要，但技术含量也无非就是测试计划—理解需求—测试用例设计—发现验证 bug—测试报告等。不专业一点讲，功能测试的技术含量就是点点点，稍微有点脑子的人都会，这也是测试行业人员相对于开发人员工资普遍偏低的原因。目前来看，企业招聘测试人员时，要求越来越倾向于会性能、会自动化、会流行语言、会 Java 语言、会 Python 语言，这是好现象，也是发展进步的必然趋势。

Web 功能自动化，对项目不太有价值，建议兼做即可，顺便提升一下开发技术并了解原理，增加与开发人员沟通的筹码，一起把项目做好。技术提升对自己是有必要且有价值的。

Android 功能自动化，对项目不太有价值，建议兼做即可，由于 android 机型款式较多，可以用自动化做些机型适配测试提升测试效率。不太可能全自动化、自动化减少多大人力、提升很大质量。技术的提升对自己是必要且有价值的。

iOS 功能自动化，对项目不太有价值 建议兼做即可。不太可能全自动化、自动化减少多大人力、提升很大质量。技术的提升对自己是必要且有价值的。

性能自动化测试可应用在有大量用户使用的系统中对项目很有价值。比如走秀网早期由于网站支撑不了，系统几乎瘫痪，导致公司受到重大损失，以至于不得不重新构建开发系统。

接口自动化测试主要在有大量接口的系统中使用，对项目较有价值，要尽量早期介入，覆盖率要 100%。越早暴露问题，越能减少后期的维护成本和沟通成本，大大提升效率。